T5-CQG-158

Fernald Library
Colby-Sawyer College
New London, New Hampshire
Presented by
MRS. FRANCES RUTTER

BRAINS AND REASONING

Brain Science as a Basis of
Applied and Pure Philosophy

By the same author

Molecular Control of Cell Differentiation and Morphogenesis

Psychobiological and Neuroscientific Theories

BRAINS AND REASONING

Brain Science as a Basis of Applied and Pure Philosophy

GERHARD D. WASSERMANN

University of Newcastle upon Tyne

ARCHON BOOKS

Library of Congress Cataloging in Publication Data

Wassermann, Gerhard D
Brains and reasoning.

Bibliography: p.
1. Neuropsychology. 2. Science - - Philosophy.
I. Title. [DNLM: 1. Neurophysiology. 2. Psychophysiology. WL102 W322b]
QP355.2.W37 501 74-14875
ISBN 0-208-01471-3

First published 1974 by
THE MACMILLAN PRESS, LTD
London and Basingstoke
and in the United States of America
as an Archon Book, an imprint of
THE SHOE STRING PRESS, INC.
Hamden, Connecticut 06514

Printed in Great Britain

To
Peter and Gisela Held
and my late teacher
Professor Alfred Menzel

PREFACE

Neuropsychology is concerned with neuroscientific mechanisms which enable brains to represent and generate the range of phenomena studied by experimental psychologists, ethologists and others (for example, psychiatrists). Neuropsychologists are repeatedly faced with some deep philosophical issues (*cf.* Eccles, 1964, 1970; Sperry, 1969, 1970). If they turn to philosophers, they may find that contemporary philosophy of science, when its foundations are followed far enough, raises many unsolved neuropsychological problems of its own, but has answered few, if any, philosophical questions of neuropsychology satisfactorily. In order to deal with this dilemma, I propose, as a *Leitmotif* of this book, that *a neuropsychology–philosophy feedback cycle can be established* which, *by successive approximations*, may provide gradually improving simultaneous answers to major problems of neuropsychology and many branches of philosophy, including the philosophy of science. The notion of a (non-circular) *scientifically founded* philosophy of science will be more fully explained in § 1.1, whose conclusions also extend to some major branches of 'pure' philosophy.

Neuropsychologists apart, many scientific theorists have taken too little notice of philosophical problems. The urge to get on with their own job may tend to prevent practising theorists from paying adequate attention to the philosophy on which their whole enterprise rests. In fact, on the basis of numerous personal discussions, I concluded that many theorists are not even aware that unresolved, serious philosophical problems remain for the whole of science. They are prepared to accept current practice in theorising, as far as their everyday work is concerned, since familiarity with the complexities of actual situations has given them an intuitive grasp of the dangers of hidden pitfalls. There are also scientists who mistakenly believe that it is a 'trivial matter' to clear up some of the more profound scientific–philosophical issues.

At the other extreme we find quite a few philosophers of science who have tried to schematise some essentials of scientific theorising (such as the procedures used in generalising and in making deductions). However, essential parts of these schematisations are often a caricature of scientific practice, even if there is a good deal of value in other aspects of this work. The philosophers' efforts have relied heavily on formal logic, which is quite unsuited to simulating the *approximate* way in which many important theories (in physics, for example) are being applied in practice. Hence, before trying to apply philosophy of science to neuropsychology, we must first examine existing major weaknesses of the philosophy of science and indicate ways in which they could be remedied. A part of this book is devoted to this critical task.

Research in various branches of theoretical physics and theoretical biology gave me an opportunity of first-hand acquaintance with the structure and applications of many simple and complex scientific theories in physics, molecular biology and neurobiology among others. I was directly faced with some of the problems discussed in this book during extensive work on a new neuropsychological theory (that is, a 'brain-model'). A few preliminary hints of this theory were included in my recent monograph (Wassermann, 1972, chapter 17), which provided also the theoretical neuroembryological foundations and genetic control theory of the new brain-model. A more extensive outline of the new ideas can be found in my second monograph (Wassermann, 1974). In the final chapter of the present book I shall attempt to separate the metaphysical aspects of the mind–body problem from legitimate aspirations of theoretical neuropsychology. The present book provides a philosophical background for my new brain-model. Yet the book is self-contained and deals also with many central aspects of the philosophy of science.

Gerhard D. Wassermann
University of Newcastle upon Tyne

CONTENTS

1 NEED FOR A NEURO-PSYCHOLOGY–PHILOSOPHY FEEDBACK CYCLE IN THE PHILOSOPHY OF SCIENCE AND IN 'PURE' PHILOSOPHY

1.1 Neuropsychology and the philosophical problems of scientific theorising

Some of this book is intended to shatter complacent beliefs that philosophers of science can, *unaided* by the use of neuropsychological hypotheses, gain a *deep* understanding of the nature of scientific theorising. I shall argue that, for a penetrating understanding, a continual feedback cycle between neuropsychology (as defined in the preface) and the philosophy of science is required, each partially providing the foundation for the other. As neuro-psychology cannot be strictly separated from other sciences, their hypotheses may *via* neuropsychology also enter the feedback cycle, thereby forming a more general feedback cycle between scientific theories and the philosophy of science. However, neuropsychology is the *primary* science involved in the feedback cycle. In fact, similar arguments could be extended to branches of 'pure' philosophy (such as epistemology). As J. S. Mill did not adequately develop his important views about the psychological foundations of logic (*cf.* Britton, 1953, pp. 142–3), his ideas were summarily dismissed as 'psychologisms', with the result that much of contemporary philosophy of science and of 'pure' philosophy became nascently condemned to rest on inadequate premises.

Mill's arguments were, in fact, only fragmentary hints, which may have contributed to their premature dismissal. By extending and modifying Mill's views, however, we can obtain a powerful new constructive method for examining old problems. Many philosophers of science seem to believe that much of scientific theorising can be adequately represented by formal logic, and that various proposed probability measures of 'confirmation' of hypotheses can be related to formal logic. However, formal logical arguments do not remotely correspond to the actual way in which many advanced deductive calculi are often used in physical theories. Complex scientific deductions (for example, in quantum mechanics, electromagnetic theory, classical mechanics, and so on) frequently rely on *approximation procedures* instead of rigorous use of deductive calculi (for example (i) perturbation theory, (ii) approximations based on the calculus of variations, (iii) use of the method of steepest descent, and so on). In addition, they may use numerical computation techniques involving sometimes accumulative errors. (I am leaving aside self-correcting successive approximation procedures, like the Fock–Hartree technique in quantum mechanics.)

To regard deductions involving approximation procedures, and possibly cumulative errors, as being equivalent to formal logical deductions is completely unwarranted. In fact, non-logical brain activity is required in order to *evaluate* the quality of an approximation, by applying various criteria which may vary from case to case. For example, to assess how well a deduction based on approximations agrees with experiments, or to compare such approximations with similar ones used in simpler cases, *decisions* have to be made by human brains, and decisions are not reducible to formal logic. In the quantitatively less extensively developed sciences (such as molecular biology), interpretations are often haunted by the spectre of experimental artefact (*cf.* Wassermann, 1972, chapter 3, for examples). It is again the interpreter's evaluative brain activity and not the application of formal logic which may lead to a *decision*, that the experimental procedure used, possibly jointly with other evidence, speaks against an artefact.

When fellow scientists inspect scientific theories and interpretations, it is the agreed judgments by many brains which may ultimately decide acceptance. We have only to follow recent assessments of discoveries in molecular biology in order to appreciate the fluctuation in appraisals, until issues become settled for the time being (although they may be reopened by new discoveries a few years later) (compare the critical weekly reviews under the headings 'News and Views' in the journal *Nature*).

The preceding arguments suggest that the use of formal logic (i) is not equivalent to *approximate* deductions derived from scientific hypotheses, which, for the purpose of making deductions, are treated as if they are (provisionally) *logically* 'true'; (ii) provides no clue concerning the validity of approximations; and (iii) does not *per se* provide evaluative decisions in the case of possible experimental artefacts. I conclude that many of the most important scientific procedures depend on brain representations which are not of the same variety as those involved in the application of formal logic. This leads to a *scientific* question: what kinds of brain representations are used in the invention, utilisation and evaluation of approximation procedures, and in the assessment of artefacts (and the like)? In fact a 'neuropsychological' *theory* which relates brain function to the brain representation of cognitive processes (and so forth) could provide a *deep* foundation for scientific theorising, for the evaluations of experiments and connected procedures. It could similarly lead to a new basis for epistemology (and so on) in 'pure' philosophy.

Prima facie, this modernised version of Mill's 'psychologism' (now replaced by a 'neuropsychologism') may appear circular. One could argue that the *philosophical* foundations of neuropsychological theories could only be 'clarified' (if at all) *via* the philosophy of science, and that, prior to its own philosophical clarification, neuropsychology could hardly be used for elucidating basic 'problems' in the philosophy of science. Circularity can, however, be avoided by adopting a procedure of *successive approximations*. In zero order approximation we may consider man as a 'black box' who, by neuroscientifically unspecified neuropsychological processes, can generate hypotheses, invent deductive calculi and approximation procedures for these calculi, and who can evaluate results obtained by these and other methods. This zero order approximation cannot attempt to *justify* anything, but merely serves to pinpoint the problems and to characterise common aspects of scientific theorising. It classifies those practices of scientific research which are widely shared by many sciences.

In the first order approximation we can no longer regard man as a black box. Instead we must propose theories which suggest how brains could execute and physically represent all kinds of cognitive processes, since these are utilised in scientific theorising. These theories, for instance, will have to explain how brains can represent concrete 'images' of visual configurations (see Wassermann, 1974) and 'translate' these, *via* motor neurons and muscles, into diagrams. Again, they must explain how brains can represent concepts, and translate these, *via* efferent systems, into physical representations of grammatically ordered, meaningful symbol sequences of spoken or written language. This is necessary because much of scientific discourse makes use of ordinary languages, and similar problems also arise for 'artificial languages' (such as mathematics). We must also explain how brains can generalise from a few inspected samples of things, events or situations, to classes which may contain an indefinite number of variants of these samples. Numerous attempts by philosophers to discuss generalisation procedures used by scientists failed because they were only carried out in the zero order approximation—that is, without attempting to explain how brains can produce scientific as well as non-scientific generalisations. In fact, the zero order approach to the philosophy of science parallels radical (as distinct from methodological) behaviourism, which attempted, unsuccessfully, to explain human and animal behaviour with complete disregard for the potential role which brain-models (that is, neuropsychological theories) could play in this connection.

A brain-model is a hypothetico-deductive neuropsychological theory, which tries to explain how brains generate representations of the essential class properties of perception, creative thinking, motivation, value formation, motor activities and much else. In particular, brain-models should try to explain how human speech activities can symbolise cognitive representations by brains. Higher order approximations to the proposed feedback cycle between brain-models and the philosophy of science can be expected as new and better brain-models are invented in the light of experimental progress. New brain-models must be scrutinised for a possible yield of better explanations of some of the processes used in scientific theorising (for example, in generalisation, hypothesis construction, decision making, and so on).

A successively better understanding of the brain processes used in cognition, learning (and so on) should help to provide a sound theoretical basis for experimental psychology, as well as explain scientifically the intellectual procedures which we use in theorising and in 'pure' philosophy. In our present (largely zero order) approximation we can theorise about brains without being able to explain *properly* how brains and their owners theorise. Only the first order approximation of the proposed feedback cycle can form a start for such an explanation. After all, we can eat without understanding the physiology of digestion. However, much human scientific effort (in, for example, cosmology and astronomy) is directed towards attaining, step by step, a better *understanding* of how things work. The preceding argument suggests that science can contribute to an understanding of its own foundations by elucidating the brain processes which man uses in acquiring scientific knowledge.

If brain-models are to cast fresh light on basic problems of the philosophy of science, then this branch of philosophy will depend on the foundation-hypotheses of brain-models. Hence, first and higher order approximations to the philosophy of science, like scientific theorising itself, cannot be hypothesis-free. This conclusion is consistent with Popper's theory of knowledge (see p. 32), according to which all knowledge depends on hypotheses, but it is at variance with Wittgenstein's view, that all philosophy should be hypothesis-free. As long as investigators adhere to the *zero order approximation* to philosophy of science, Wittgenstein's notions are adequate, but if they wish to attain a deeper grasp of the problems involved, then, I believe, Popper's orientation is more appropriate. (I am here referring only to Popper's theory of knowledge, as I do not endorse all of his views.)

Various philosophers (such as Britton, 1953, pp. 142–3, and Feigl, 1958, p. 418) disapproved of John Stuart Mill's disinclination to separate formal logical thinking from other brain-produced cognitive (or 'psychological') processes. According to Feigl (1958, p. 418), Mill 'regarded logical truths as on a par with the truths of the natural sciences' and, in Feigl's opinion, Husserl and Frege initiated a phase of 'clarification' which put an end to this supposed error on the part of Mill. However, the arguments of § 1.3 below imply that Mill's point of view, when presented in a new guise, is more plausible than that of his critics.

Mill's critics disregarded the fact that formal (for example, mathematical or logical) arguments and empirical arguments *seem* to differ *explicitly*, simply because many philosophers have adopted an arbitrary system of classification which makes them differ. In fact, I shall argue that formal logical arguments, as well as empirically-bound thinking (such as everyday thinking or scientific arguments), *all* depend on brain-representations which have *empirical* antecedents. Thus, at a deeper level of enquiry, the difference between formal and empirical statements vanishes (see § 1.3), even if it persists at the shallow level of classification. Some of Mill's critics implicitly regarded Wittgenstein's 'zero order approximation' to philosophical problems (see p. 4) as if this had come to stay, instead of recognizing (like Gellner, 1959) its sterility. They did not anticipate that 'higher order approximations' to philosophical problems (in the sense of pp. 3–4) could be possible, which abandon the Wittgenstein ideal of a hypothesis-free philosophy and, by introducing a feedback cycle between neuropsychology and the philosophy of science, revitalise the latter *and philosophy in general*. (For a specific critical evaluation of the views of a variety of 'linguistic philosophers' the reader is referred to Mundle (1970).)

1.2 Artificial attempts to separate philosophical and scientific issues

A. Contemporary zero order uncoupling of science and philosophy

Many philosophers of science who are influenced by the Wittgenstein tradition believe that they should expound the typical features of *existing* scientific methodology (such as typical aspects of procedures of theory construction), and that they should discuss the 'problems' which this methodology raises. It is fashionable to assume that many philosophical problems are essentially 'logical' ones, even if some philosophers concede that for a variety of reasons present forms of logic may have to be replaced by more appropriate ones (Körner, 1966; Finkelstein, 1969; Putnam, 1969). Their criteria of appropriateness differ from my own, which assert that, in discussing scientific theories, formal logic is, in most cases, essentially inappropriate, because the results of approximation procedures used (in deductive procedures, for example) are not equivalent to formal logical implications (see pp. 1–2).

True enough, sometimes deductive *formalisms* (plus interpretations) of theories, such as quantum mechanics, are *in parts* precise *per se*, and can be related to formal logic (*cf.* Putnam, 1969). However, such formalisms can only be *applied* rigorously to empirical systems in a few textbook cases. Many steps which relate theories to experiments involve, as a rule, not only the use of *ad hoc* simplifications of theories, but also approximations which cannot be logically justified. The clarificatory (or 'justificatory') problems of theory construction and of other aspects of philosophy of science (and of 'pure' philosophy) are therefore not equivalent to logical issues. They involve additional cognitive problems (such as valuations or decisions). I emphasised that possible solutions to these problems can be gradually approached by a series of successive approximation cycles, involving a feedback between neuropsychological theories (brain-models) and philosophical issues in science, epistemology and related topics (see § 1.1).

In the present state of the art, where the zero order approximation is exclusively in use, there exists a complete decoupling between neuropsychology and philosophy (and the philosophy of science). This is clearly illustrated by Sir Alfred Ayer's monograph (1956), which contains chapters headed 'perception' and 'memory'. On comparing the contents and arguments of these chapters with *scientific* monographs or papers (a) on perception (such as Teuber, 1960; Hochberg, 1964, 1971; Zusne, 1970; Vernon, 1952, 1962, 1970; Rock, 1966) or (b) on memory and learning (such as Hilgard and Marquis, 1961; Kling and Riggs, 1971), we realise that Ayer does not include discussions of any relevant scientific aspects of experimental psychology or of the neurosciences. On the contrary, he aims to convince his readers that philosophers can solve their problems *without* resort to science of any kind.

In keeping with this, Ayer (1956, p. 7) asserts that 'philosophical theories are not tested by observation. They are neutral with respect to particular matters of fact.' The view that philosophy is independent of *specific* empirical information is contained in Ayer's (1956, p. 7) claim that philosophers

> are in the strange position that all the evidence which bears upon their problems is already available to them. It is not further scientific information that is needed to decide such philosophical questions as whether the material world is real, whether objects continue to exist at times when they are not perceived, whether other human beings are conscious in the same sense as one is oneself. These are not questions that can be settled by experiment, since the way in which they are answered itself determines how the result of any experiment is to be interpreted. What is in dispute in such cases is not whether, in a given set of circumstances, this or that event will happen, but rather how anything at all that happens is to be described.

Ayer chose illustrations that, surprisingly, bear out the weakness of his case, which rests on the contemporary uncoupling of science and philosophy. According to § 1.1. this can only be remedied by a scientific–philosophical feedback. For some philosophers, 'uncoupling' provides a source of comfort, furnishing an excellent excuse for staying clear of science. In fact, I shall argue that (i) the issue of the reality of the material world is undecidable at any level of approximation of the science–philosophy feedback cycle; and (ii) the issue of the existence of objects (such as postulated particles of Newtonian mechanics) when not observed is in many, if not all, cases an empirically linked issue.

For example, classical mechanics assumes that forces act on particles while not observed, and hence that the particles must continue to exist during that period. Without these assumptions classical mechanics would not be valid. As a second example, consider studies of enzymatic reaction kinetics, where organic molecules may be rapidly assembled, changed and/or dissociated (depending on the case). Existing theories assume explicitly that (postulated) molecules of certain types may, in certain reactions, cease to exist (in their original form) during periods when a reaction system is not being observed. More generally, the existence of objects in states which are no longer observable (such as the past history of a chemical solution or the past history of a pulsar) is often indirectly inferred from scientific theories whose premises (that is, initial hypotheses) are indirectly related to independent empirical evidence.

Further, (iii) the question whether other people are conscious in the same sense as I am must also be considered in the light of empirical issues (see my arguments relating to Feigl's discussion, pp. 43f below).

B. *Different designations of the word 'theory'*

A difficulty arises from the fact that the word 'theory' means different things to different people. Pure mathematicians may speak of 'the theory of Hilbert space'. While this entirely formal 'theory' has a deductive structure, it is, by classification, not a *scientific* theory since it is devoid of (explicit) empirical reference (see p. 4). Again, when some philosophers speak about 'philosophical theories', it is apparent from accompanying statements that their 'theories' have nothing in common with scientific theories (compare Ayer's use of the word 'theory' on p. 6 above).

C. *Decision procedures*

When we are confronted with either scientific or philosophical problems, a fundamental question arises: what methods exist for *deciding* the validity of answers to these problems? In a related context, Minsky (1959) wrote: 'How can we decide when a question has been properly answered? Clearly this question itself is not well-defined until we have *chosen* (not discovered) its answer.' In the sciences, answers to many questions are given in the form of 'explanations' of properties of systems or events. The *choice* of answers in this case is provided by formulating *hypotheses* and accepting them at least temporarily. Hypotheses appear often interlinked, forming a 'network', each hypothesis of the network being directly or *via* others (in exact or approximate form) 'empirically anchored', that is, related to publicly accessible observations. An empirically anchored network of hypotheses is called a *scientific theory*. This crude preliminary description will later be amplified by many additions and qualifying statements.

Depending on the case, the direct or indirect empirical anchorage of each hypothesis of a scientific theory allows either (a) individual hypotheses, or (b) subsystems of interlocked hypotheses, or (c) the complete set of hypotheses of the system to be subjected to a variety of *experimental tests* which may help to confirm or refute the plausibility of either some or the combined system of hypotheses of the theory. Methodologically, experiments may be regarded as *part* of a decision procedure for establishing, at least on a temporary basis, the acceptability of one or more hypotheses of a scientific theory (that is, of deciding the plausibility of answers to scientific questions in the light of the empirical evidence available at the time). In fact, many scientists do not claim the 'truth' of their hypotheses (except, sometimes, in a formal logical sense; see p. 2). Likewise many scientists, while talking for the sake of convenience *as if* certain entities are real, do not intend to imply the reality status of certain entities (such as elementary particles). Following general practice, they will only claim that certain hypotheses (referring to *properties* of 'entities') are well supported by empirical evidence, *for the time being*, but that, in the light of new empirical evidence, some, but not necessarily all, of their hypotheses may have to be replaced or altered.

The passage cited from Ayer (p. 6) illustrates the types of questions that certain contemporary philosophers pose. If we ask whether their answers have been properly chosen, it is found that philosophy, when decoupled from science, lacks a systematic decision procedure for evaluating the validity of its assumptions (that is, of its *chosen* answers in the sense of Minsky—see above, and see Bunge, 1967c, vol. 1, p. 216). There are no valid reasons for accepting Ayer's arguments that the interpretation of the result of any experiment depends on preliminary answers to philosophical questions (for instance, of the type which he gave as examples). No evidence exists that accepted interpretations of results in (for example) molecular biology depend on the hypothesis (i) that 'other human beings are conscious in the same sense as one is oneself' (*cf.* § 2.1B), or (ii) that 'the material world is real' (see p. 6 above).

Doubts arising from philosophical scepticism *per se* do not affect decisions concerning the acceptability of scientific theories. The reason for this is the attitude of most scientists. Their accepted theories rest on initial hypotheses, which most scientists regard as 'plausible' or 'good bets' (until other evidence may cause a revision of opinions). Theories establish a great deal of order among observed phenomena by relating superficially unconnected phenomena with one another at a deeper level of theorising.

The thesis of § 1.1 leads to the conclusion that those philosophical 'problems' whose answers cannot be decided by a feedback between appropriate scientific theories and philosophy are purely metaphysical: they can be discussed only on the basis of arbitrary personal opinions (like beliefs of mystics or art criticisms). In contrast to this, presumptive decidable philosophical problems can, in my opinion, be approached only by the feedback cycle linking philosophy with neuropsychology (and other sciences; see p. 1). Even then, however, any 'acceptable' decision (reached at some stage of the cycle) rests on the acceptance of scientific hypotheses, which may always be subject to further revisions. Accordingly, final and certain answers are never possible either in science or philosophy, even at the deepest levels.

D. Undecidability of 'reality' problems

Suppose that we wish to decide whether philosophical questions concerning the 'reality' of the 'material world' have been properly answered (*cf.* p. 6); then we ought to examine (according to Minsky's dictum) how philosophers *choose* their answers. We must ask: what do people mean by the 'material world'? The answer depends on the *level* of discourse adopted (see § 1.5F). Scientists, unlike the man in the street, suggest limited but often highly sophisticated answers in terms of *theories* concerning the hypothesised structure and behaviour of certain classes of 'systems' (such as liquids, gases or molecules). The hypotheses relating to these 'systems' provide the *relevant* aspects of the 'material world' as far as *scientists* are concerned.

As scientific hypotheses about the 'material world' are often rapidly changing with improvements of experimental techniques (compare the earlier ups and downs concerning evidence for the 'existence' of reverse transcriptase in molecular biology—see Temin and Mizutani, 1970; Ficq and Brachet, 1971), there exists no unchanging meaning of (theory-bound) descriptions of the 'material world' (see also Mercier, 1970). Attempting to decide the reality status of physical 'systems' (such as electrons) which, as time passes by, may be associated with totally different theoretical descriptions is like trying to decide whether a criminal who has altered his appearance and fingerprints by skin grafts, and changed his name, is the 'same real man' he was before these changes took place. Answers to such questions *cannot be* decided by any conceivable means, since the questions themselves cannot be unambigously formulated. Some philosophers are inclined to escape from this dilemma by claiming that the 'real world' to which they refer is not that described by scientists and embodied in scientific hypotheses, but the world described by the language of 'the man in the street'. This method, however (even if it were fruitful, which is far from evident), could not be extended to the realm of science which is not encompassed by the discourse of the 'man in the street'.

Inspection of a typical sample shows what one representative philosopher understands by 'existence' (that is, 'reality') of physical objects. After a lengthy discussion of this topic, Ayer (1956, p. 132) claimed that:

> In the end, therefore, we are brought to the unremarkable conclusion that the reason why our sense-experiences afford us grounds for believing in the existence of physical objects is simply that sentences which are taken as referring to physical objects are used in such a way that our *having the appropriate experiences counts in favour of their truth*. It is characteristic of what is meant by such a sentence as 'there is a cigarette case on this table' that my having just the experience that I am having is evidence for the truth of the statement which it expresses. The sceptic is indeed right in his insistence that there is a gap to be overcome, in the sense that my having just this experience is consistent with the statement's being false; and he is right in denying that a statement of this kind can be reduced to a set of statements about one's sense-experiences, that is, to a set of statements about the way that things would seem. He is wrong only in inferring from this that we cannot have any justification for it. For if such a statement functions as part of a theory which accounts for our experiences, it must be possible for them to justify it. The very significance of the theory consists in the fact that its statements can in this way be justified. [Italics are mine; see also my comments in § 2.2.]

As Ayer does not expound the detailed structure of the 'theory' to which he refers, it is not possible to *decide* whether his 'answer' concerning the existence of physical objects is valid. In fact, in a passage preceding the one already quoted, Ayer (1956, p. 132) asserts that:

> . . . in referring as we do to physical objects we are elaborating a theory with respect to the evidence of our senses. The statements which belong to the theory transcend their evidence in the sense that they are not merely re-descriptions of it. The theory is richer than anything that could be yielded by an attempt to reformulate it at the sensory level. But this does not mean that it has any other supply of wealth than the phenomena over which it ranges. It is because of this, indeed, that they can constitute its justification. Accordingly, it does not greatly matter whether we say that the objects which figure in it are theoretical constructions or whether, in line with common sense, we prefer to say that they are independently real. The ground for saying that they are *not* constructions is that the reference to them cannot be eliminated in favour of reference to sense-data. The ground for saying that they *are* constructions is that it is only through their relationship to our sense-experiences that a meaning is given to what we say about them. They are in any case real in the sense that statements which affirm or imply their existence are very frequently true.

The last sentence represents a zero order approximation to a philosophical issue (in the sense of p. 3). The decision about reality is to be based on the criterion of 'frequent truth' of a 'statement'. The 'truth' referred to is certainly not that of formal logic. In fact the word 'truth', as used in Ayer's sentence, refers to an 'existence' statement. As he provides no *independent* criteria for deciding in which sense an object 'exists', apart from language use of the word 'existence', we cannot say that any statements implying existence of objects are 'true'. Hence, the validity of Ayer's zero order approximation to 'reality' or 'existence' is undecidable, by any existing criteria.

Apart from this it is not self-evident that, as suggested by Ayer, *theoretical constructs* can be granted reality status. On the contrary, I shall argue later that their reality status is also undecidable. I shall suggest below (1) that 'reality' has no absolute status, but is related to neuropsychology; (2) that its bearing on philosophy arises *via* the proposed neuropsychology–philosophy feedback cycle of § 1.1; and (3) that the 'reality' discussion of Ayer proceeds at a different level from that adopted by philosophers of science.

Ayer is only one of many who have discussed the problem of 'reality'. Its meaning in physics and elsewhere has been widely considered (*cf.* Yourgrau, 1964, p. 362; Moore, 1922, 1953, 1959; Quine, 1953, 1961). Yourgrau emphasised that Moore and Quine 'associated "reality" with respective attitudes adopted by workers in various domains of inquiry'. Any claim of 'absolute' reality status of *hypothetical* physical systems (such as atoms or electrons) is metaphysical (see p. 106), owing to lack of decision criteria. It possibly represents an extrapolation from brain-generated 'reality' of immediately perceived objects and movements (*cf.* Vernon, 1952, p. 191). If this is correct, then reality 'properties' could be among the representations of *values* which brains form, and associate with brain representations of perceived things, situations and events.

In this sense, 'reality' could have a status comparable to other brain-represented values such as 'beauty' or 'ugliness'. Just as the assignment by man of reality status to immediately perceived objects, such as trees, houses or people, depends on the function of human brains, so there exist no *absolute*, but only brain-created, criteria for judging values (such as whether a picture is beautiful). Although an individual brain will associate its representations of specific values with its representations of specific things, such associations could vary widely among people (some people prefer pop music to Bach).

Successively better brain-models should (i) provide gradually improving explanations of how brains represent values, and (ii) associate these values (in some cases probably by innate mechanisms) with engrams (that is, brain-memory-traces) of objects, events, concepts, and so on. (For instance, a particular combination of detectable subengrams—such as a word in a remembered sentence—could be associated with a particular brain represented value.[1]) Assignment of reality status, as well as of values, to brain representations of objects, situations, events and concepts could then involve similar types of brain processes. Yet a brain assignment of 'reality' status to objects (and so on) does not imply that objects are real in any absolute sense, any more than a brain assignment of a value (say, 'beautiful') to an object implies that the object has this value in any absolute sense. Beauty lies in the eyes of the beholder!

This demonstrates in an explicit manner how brain-models could help to *explain* the human tendencies to perceive certain objects as real (just as they perceive things as being 'beautiful'), and then extrapolate this perceptually formed association to situations where the perceptual machinery cannot come into play (for example, when the reality of 'electrons' and other theoretical 'systems' is being discussed). Philosophers, unaided by science, could not explain the manner in which man (*via* his brain mechanisms) (i) evaluates *hypotheses* referring to electrons, or (ii) associates direct reality *values* with trees but not with electrons, because cues referring to electrons are not evaluated by brains in the way in which houses and trees are perceived.

Even at the level of everyday discourse there are difficulties for those 'linguistic philosophers' who believe that such discourse can decide the reality of objects. When people converse with one another, they frequently refer to objects, people, and so on. Yet the objects (and so on) referred to may

[1] My recently developed brain-model, the beginnings of which have already been outlined (Wassermann, 1974), relies on molecular mappings, and particular values could be physically represented by specific 'mapping molecules' (proteins) of particular classes of molecular maps, located on synapses, as explained in Wassermann (1974).

only exist as memory-traces (engrams) in people's brains, and this does not imply the existence of these objects (and so on). Memory distortions suggest that engrams, or their reactivation facilities, can change. Hence the reality status of past events could not be linked, reliably, to memories. *The fact that normal people understand one or more languages depends, for instance, on the formation of associations between language engrams and engrams of objects* (which are not the same as the objects themselves). It certainly suggests that different (normal) people can form 'similar' brain representations of similar objects, language symbols, and so on, as otherwise many words could not be similarly responded to.

However, experimental psychology shows that there exist detectable differences in the perceptual functions of different people (and animals), depending on attitudes and presumably on genetic factors (among others). Hence, in what sense brain representations of the same thing are 'similar' for different people cannot generally be known, except (conceivably) in a few rigidly controlled experiments. As brains represent objects only indirectly, and as language at most refers to brain representations of objects (and so on) (in our own brains or those of other people), language does not refer to objects (and so on) themselves. It is also for this reason that allegedly 'true' statements of language cannot imply the reality of objects *per se* (*pace* Ayer's statement on p. 10). In fact, many experiments imply that brain representations of visual, haptic (or other) configurations do not *uniquely* correspond to the things represented—for example, in visual geometrical illusions the same straight line on paper may appear longer or shorter, straight or bent, depending on the nature of other neighbouring contours; again, attitudes can affect size constancies and so on.

I conclude that we can have no reliable criteria for deciding what we mean by the reality of objects, events (and so on). Indirect evidence also indicates that brain representations of language and of patterns (involved in pattern recognition) can be dissociated, which implicates the independence of these representations. Their normally present association is consistent with association errors in case studies of agnosia. For instance Geschwind's patients (1969, pp. 117–18) made naming errors of objects or failed to give correct responses when offered a multi-choice list which included the name of a visually presented object. This shows the important part played by brains in associating their representations of language symbols with the representations of things, events, concepts, and so on, which the language expresses. While discussing visual agnosia, Geschwind referred to a few typical publications (Lissauer, 1889; Lange, 1936; Brain, 1941; Macrae and Trolle, 1956; Ettlinger and Wyke, 1961; Hécaen and Angelergues, 1963) which provide 'frequent evidence of preservation of non-verbal visual identification while verbal identification is impaired', suggesting dissociability of the brain's language modality from pattern recognition. (The reader is referred to Geschwind's interesting comments.)

The facts just mentioned reinforce the view that language statements do not entail any absolute 'reality' of perceived objects (and so on), but indicate strongly that these statements are related to brain representations of objects (and so on). Epistemologists have too often been concerned with the relation between private experiences and the reality of events or objects (and so on) in the world (now or in the past). Empirical tests can at best help to confirm or refute ('falsify') the various hypotheses of scientific theories, but hypotheses are only conjectures about objects, events (and so on), and are never 'true' in any absolute sense. Moreover, private experiences are publicly inaccessible and hence also have the status of hypothetical systems. As they are 'private', they are neither directly nor indirectly (*via* scientific hypotheses) empirically anchored (*pace* Eccles, 1964, p. 271). (Many scientists who use behaviour symptoms such as language symbols as indicators of private experiences have realised that these symptoms are not in any definable sense *equivalent* to the private experiences which they are *assumed* to represent in publicaly accessible form. Reasons for this will be given in § 2.1. If a subject whose leg has been amputated experiences a 'phantom limb' phenomenon, it is his public statements that scientists and others can examine, but not his privately experienced sensations relating to the region of the missing limb.)

1.3 The empirical bondage of all statements

Philosophers and other find it useful to classify statements (compare Kant's classifications discussed by Körner, 1955, p. 19ff). But classifications which are performed by human brains represent empirical *operations* carried out on processed stimulus configurations provided by printed or spoken or other material. For instance, a picture can be classified as being 'abstract'. But this classification is only made by a human brain operating on the sensory input received from a very material system (in the sense of p. 9), such as a coherent aggregate of ordered paint molecules attached to a canvas or other background. Classifications, though useful for many purposes, are always arbitrary, and an object or a statement can be classified according to different criteria, which focus on different properties.

One widely adopted classification of certain types of statements is based on a division into (a) *formal* statements which include among others all formulae of logic and pure mathematics (provided these formulae are deemed valid by the classifier), and (b) *empirical* statements (*cf.* Bunge, 1967c, vol. 1, p. 22), the latter being related to facts. Empirical statements and the facts to which they refer must not be confused, as even empirical statements could contain hidden hypotheses (such as the hypothesis that a letter bearing the signature of Louis XV was really signed by that king).

Formal statements are supposed to be logically 'true'—that is, their 'truth' is supposed to be independent of the existence of empirical systems, even if the statements contain *references* to empirical systems. (These statements are also known as *a priori* statements, and some philosophers thought that all philosophical propositions ought to be *a priori* statements (e.g. Russell, 1953, p. 107). Russell (1953) argued that a 'philosophical proposition must be such as can be neither proved nor disproved by empirical evidence'. This is equivalent to saying that it must be logically 'true'.) Körner (1955, p. 19) exemplified this concept of *a priori* statement with reference to Kant, in the following passage:

> Now if a judgement is to be *a priori* it must be logically independent of all judgements which describe experiences or even expressions of sense. Examples are the judgement that 2 + 2 = 4, that an equiangular Euclidean triangle is necessarily equilateral, that every father is necessarily male. It is of course true that these judgements have a kind of dependence on experience. We form them as a result of certain experiences and in the course of reacting to them. However, this dependence is not what is meant by logical dependence. The above *a priori* judgements would be equally true in a world having no separate countable objects, no shapes, and no differences of sex.

The notion that formal (or *a priori*) statements are logically 'true', independently of the material world and its systems, is not only inherent in the last sentence of Körner's passage, but is widely accepted by certain philosophers. For instance, many people accept the logical 'truth' of Boolean algebraic deductions, on the assumptions of logically 'true' or 'false' starting formulae, provided reasons for the consistency of the system can be given. But they disregard the fact that these deductions are made by human brains or computers, that is, by material systems. Hence the demonstration of the truth of these deductions is *not* independent of systems of the material world. (This aspect of the status of formal logic is left aside by those who question whether existing forms of logic are appropriate as deductive calculi when applied to empirical investigations (e.g. Putnam, 1969).)

I fail to see how the logical 'truth' of the statements contained in the passage cited from Körner could be independent of empirical systems. To evaluate the logical 'truth' of these statements requires empirical systems, such as brains or computers. In other words, the world of material things is essential for the validation of logical 'truths'. The statement that 'all fathers are male' only becomes a logical 'truth' (a tautology in this case) if one defines *via* a dictionary (which is also a material system) the word 'father' as a 'male parent'. But even then, as far as we know, only human brains, or equivalent material systems, can detect that the dictionary definition necessarily implies maleness of the father. Small children, who have never consulted a dictionary, learn to *associate* the word father with 'male', because their brains produce this association. Likewise the concept '+', which appears in the equation 2 + 2 = 4, only 'makes sense' to brains and computers.

Just as the material composition of a picture makes it an empirical system even if its subject matter is abstract, so the possible 'truth' of a formal statement does not make the decidability of this 'truth' independent of empirical systems. Hence even formal statements are empirically bound—that is, the decision regarding their logical 'truth' depends as much on empirical systems as the establishment of empirical statements. Before I enlarge on this point, I must briefly refer to the use of formal statements in science. When formal statements are applied in the sciences, they often acquire explicit empirical significance by becoming turned into scientific hypotheses (such as the 'laws' of ideal gases). This will be discussed later in detail. Yet these hypotheses, even if consistent with facts in the light of certain widely accepted decision procedures (such as least square curve fittings), can never be *proved* to be true by any finite number of facts. Hence, although many scientific hypotheses are formal statements, some of whose symbols are being given empirical names, the named symbols themselves are not being identified with facts but are only *related* to fact-representing symbols by very indirect procedures. For example, in Ohm's law, the current and voltage symbols which appear in the formal expression refer, not to data, but (i) to *mean values* extrapolated from data, and (ii) to hypotheses related to curve fitting (see § 2.4A).

After this detour I shall now return to the status of empirical and formal statements.

It should be remembered that all statements *per se* are meaningless. They only become meaningful when operated on by appropriate material systems, namely brains (and possibly artificial intelligence systems in the future) which can represent, classify (that is, recognise), evaluate and utilise these statements. Patterns, including the semantic contents of sentences, can be recognised only by appropriate classes of pattern recognisers. In fact, some animals cannot recognise (classify) patterns which other animals or men can discriminate and classify (*cf.* Teuber, 1960). Similar remarks apply to different classes of man-invented pattern-recognising automata (*cf.* Zusne, 1970, for extensive references). The fact that only some classes of human beings can understand certain languages, as well as many other data, implies that human pattern recognition depends in many or all cases on learning processes, which involve brain mechanisms. (Questions of innate pattern recognition capacity, based on purely developmentally established mechanisms, have been partly resolved by ethologists; *cf.* Hinde, 1966.)

Again, children can only at certain stages of brain development *understand* (recognise) certain types of relationships (such as 'identities', 'equivalence', topological and order relationships such as 'included in', 'larger than', and so on). Hence logical relations, as well as the language of 'the man in the street', are only 'meaningful' (*recognisable*) when operated on by appropriate material systems. In fact, only brains (or possibly artificial intelligence machinery) can decide the validity (or otherwise) of statements such as $2 + 2 = 4$ or $2 + 2 = 5$, and deduce the relation $(a + b)^2 = a^2 + 2ab + b^2$ from the axioms of 'real number algebra'. The ability to make mathematical (including formal logical) deductions depends therefore on physical systems, that is, brains or man-designed 'artificial intelligences'. To the best of my knowledge no philosopher has suggested criteria which could establish the logical 'truth' of the equation $(a + b)^2 = a^2 + 2ab + b^2$ independently of *any* physical system. (See also § 4.6.)

The preceding arguments do not deny the capacity of brains to allocate formal and empirical statements to different classes. But they stress the brain-dependence of these *classifications*. As most philosophers adhered to a decoupling between neuropsychology and philosophy (see § 1.1) they paid no attention to brain functions, which had serious consequences. It induced people to regard formal statements as being independently 'true' of empirical systems (that is, brains; see p. 13). This led to the unjustified belief that formal and empirical statements are *fundamentally* different, the former being ultimately empirically independent, in contrast to the latter. Already, however, in the first order approximation of the neuropsychology–philosophy feedback cycle, any intrinsic difference between formal and empirical statements appears to evaporate.

In fact, at the neuropsychological level, recognition (classification) of both formal and empirical statements depends on previously formed memory traces (engrams). Man is not born with a knowledge of typical formal statements. This has to be acquired by learning processes involving specific interactions with the environment (such as inspection of written mathematical formulae on paper, or working out of examples). (In fact, like everyday planning, or any other form of creative activity, mathematical thinking may involve a good deal of trial-and-error activity carried out by means of brain *representations*. The trial-and-error activities could be entirely carried out in terms of brain representations, only the end result being externalised by symbolic motor behaviour.)

Hence, as far as brain representations are concerned, formal as well as empirical statements depend on *acquired engrams*, that is, on empirical information. In the case of mathematics the engrams could represent (i) concrete structures (such as brain representations of geometrical figures, or of numbers), (ii) concepts (such as algebraic variables). Or there could be engram *sequences* representing rules of manipulation (for example, in algebra) or rules of order (for example, sequences of inclusions, sequences in which each algebraic quantity precedes another one) and so forth. Apart from this, *specific* engrams could induce *generalised* brain representations.

For instance, the formal statement, expressed by the algebraic formula $a + 2a = 3a$, for *any* real number a, represents a generalisation from particular engrams to rule-forming brain representations which involve a concept (namely, 'any real number'). The capacity to generalise, from engrams of concrete samples to engrams of concepts and to rules involving concepts, operates equally for brain representations of everyday objects and situations (see the example cited below). Innumerable rules known to us (such as that touching a flame can burn one's skin) have their origin in the capacity of brains to generalise from a few specific samples to classes of which the samples are members.

The heritage of Plato's idealism, with its implied absolute existence of 'universal ideas', independently of brains, led many people (such as Frege and Husserl; see p. 4 and p. 14) to believe that there exist formal relations which are 'true' independently of the existence of human brains. Unfortunately those who continue to think in the Platonic tradition have suggested no tests for demonstrating the validity of formal relations *independently* of human brains (or man-made artificial intelligences). There exists now much evidence that the capacity to generalise (for example, to form hierarchies of concepts and relations) depends in the case of man and animals on nervous classifying machinery, with different animals having different classificatory abilities.

I emphasised that recorded statements *per se* have no significance unless presented to and utilised by brains (or other appropriate physical systems). In fact, the brains of archaeologists often have severe difficulties in deciphering recorded symbolisms. While significant to the brains of long deceased societies, the concepts to which these symbols refer can be rediscovered only when contemporary brains (possibly aided by machines) detect the correct relations between symbols and concepts. Again, unlike some philosophers, I am not suggesting that statements *per se* are equivalent to brain states of their producers (see also Popper's (1970, p. 12) relevant discussion).

Consider the work of a composer. First his brain states represent the final composition which may be attained by a complex successive correction procedure. This is then translated *via* a cerebral motor representation into a symbolic motor performance, which expresses the original brain representation in the form of a musical score. This, in turn, can act on the nervous perceptual and thought mechanisms of the performers and conductor of an orchestra, and *via* their nervous motor representations leads to manipulations of musical instruments which represent their 'interpretations' of the composition.

Thus, the composer's original brain-representations can generate a multitude of other representations, some extrinsic to human brains (such as the score on paper), others evoked in other human brains. Also, man-created representations of concrete or abstract 'ideas' can either be stored by brains or be processed and transferred to man-made storage systems (such as books) for future use by other brains.

The fact that logicians or mathematicians agree to the validity of a mathematical theorem need only imply that their brains operate in similar fashion on, for example, printed records, and that the *structure* of the records must be of a particular kind to evoke similar brain reactions in different individuals. The *structure* of statements or other human records (just as the structure of everyday objects) is highly relevant for correct recognition and evaluation. But records *per se* are cognitively as neutral as icebergs and trees.

The power of human brains to classify things and relations exhibits a remarkable flexibility which is also relevant to the discussions of § 1.4. It manifests itself in a capacity to generalise from one or a few samples of a class to a 'continuum' of members of the class and to form concepts,[2] as is illustrated in the following passage from Brain (1951, p. 23).

> When someone speaks and another person listens it may seem a very simple process, but in fact, it is so extremely complicated that it is very difficult to understand. Let us take as an example the word 'dog'. No two people pronounce the word 'dog' in exactly the same way, yet we always know what it means. Not only that, it can be sung, shouted or whispered and it still conveys the same thing.
>
> . . . But we can go further than that. The written word 'dog' will still mean the same thing, though the word no longer consists of sounds, but is made up of black marks on a white piece of paper; it can still be understood, whether it is written or printed, in large or small letters, in black or coloured type, and in any sort of handwriting short of complete illegibility: in many handwritings the marks that people make on paper have very little resemblance to the letters they are supposed to represent. Here, then, there is as great a variety among the visual patterns presented to the nervous system as among spoken words. . . .

Brain mentions that the sense of touch provides similar generalising capacities in pattern recognition. 'The reader of braille can recognize a series of raised pimples on the paper, which make a pattern quite unlike ordinary letters, and yet also mean the word "dog".'

Man can also generalise previously encountered specific examples of linearly ordered grammatical and semantically correct language forms. This, presumably, requires generation of brain representations of generalised class forms, as is suggested by the large scale appearance of these forms in languages. For instance, the sentences

This house is very beautiful

Your garden is quite large

John's car is extremely small

all have the generalised class structure

$$\text{A.B.C.D.E} \qquad (1.3.1)$$

where A represents a class of 'identifier expressions' (including 'this', 'that', 'John's') which are semantically related to members of a class of subjects B ('house', 'garden', 'car', and so on). Similar remarks apply to classes C, D and E.

[2] See also Bernays' (1964, p. 38) remarks concerning the essential equivalence of concepts and the philosophical term 'universals'.

Once a human brain has established numerous class forms, any semantically compatible members of appropriate classes can be *substituted* into the correct class-form positions, thereby generating an indefinite number of variant members of each class-form. (Similarly, an indefinite number of mathematical identities or inequalities can be generated from existing ones by substituting new and more complex expressions for existing ones.) The preceding examples do not remotely exhaust the human capacity for generalisation.

1.4 Reexamination of presumptive limitations of scientific theories

I shall now examine some arguments from Körner's (1966) stimulating book. He believes that scientific theories are from the outset limited by some general constraints. However, closer inspection suggests that almost all these constraints are a consequence of a (zero order) decoupling of philosophy and neuropsychology. Most of the constraints seem to disappear when the human brain's capacities for generalisation and dynamic transformation of its representations (and so on) are taken into account. For instance, Körner (1966, pp. 91ff) expressed the view that:

> A hypothetico-deductive, or for that matter any formulated theoretical system not only suspends historical change by codifying a momentary phase of scientific theorizing; it also replaces the temporal world, which is its subject-matter, by a timeless structure of relations. This 'spatialization of time', as Bergson called it, might perhaps better have been called the 'detemporalization of change'. It is something characteristic of all conceptual thought, which turns successive temporal phases into atemporal relata. Yet this transformation may be more or less faithful to experience.

This detemporalisation occurs not only at the conceptual level of thinking but at all levels of engram formation, irrespective of whether a sequence of engrams represents a sequence of concepts or a sequence of concrete configurations. Körner has aptly implied the human brain's capacity to convert perceived dynamic events into detemporalised representations. (According to my new brain-model (Wassermann, 1974, and in preparation), detemporalised brain representations exist, in the form of static engram sequences.) However, Körner did not take into account evidence for the existence of a reverse process. This occurs, for instance, during retrieval of memories and during thought processes and involves the reactivation of engrams in correct serial order, and the resynthesis of a 'continuous' representation from presumptive static engram sequences. Reciting a poem or playing a piano score 'from memory' requires such precisely serially ordered reactivations of engrams. Likewise, an eyewitness's account of an event demands that the witness's brain can reactivate the appropriate engram sequence in correct serial order.

The perception of apparent cutaneous and visual movements, when man is presented with appropriate sequences of static stimuli (see Kling and Riggs, 1971—accounts of phi-phenomena), independently suggest that brains have the capacity to 'retemporalise', that is, to generate dynamic out of static representations. For instance, when viewing a moving film, a sequence of 'detemporalised' static frames of the film are, *via* a screen, presented in rapid succession to the retinae and from there to the brain, where (as judged by evidence; see Hochberg, 1971, p. 527) a representation of *continuously* moving configurations is being synthesised. Time-lapse films of biological systems provide particularly striking illustrations of the value of the human brain's capacity to convert static into dynamic representations. In all these cases the brain 'retemporalises' the detemporalised static frame sequence. While in the case of phi-phenomena retemporalisation processes operate on externally induced brain representations, similar retemporalisations could operate on internally stored engrams.[3]

When he reads in a newspaper or book the statically printed account of some fearful murder, the reader's 'imagination' comes into play. This could mean that his brain can generate appropriate dynamic pictorial (that is, configurational) *representations* (in mapped form; *cf.* Wassermann, 1974) of the events to which the static conceptual language symbols refer. A corresponding retemporalisation from static symbols to 'dynamic' imaginative representations could take place when a theorist inspects a printed theory or listens to an account of it. The printed 'atemporal relata' of the theory could in the reader's brain induce a *dynamic representation*. For instance, when a theorist reads about particularly shaped liquid flow-lines, he may be able to 'picture' them, even if they are not illustrated in the theoretical paper. The first limitation of theories to which Körner refers, I suggest, does not exist, because human brains can recreate dynamic representations from static representations (and from static engrams). The fact that some scientists like to think 'pictorially', others 'abstractly', about the same scientific theories suggests that the scientific theories *per se* do not force such modes of representations, but that the latter are the scientist's *brain creations*. For instance, when reading about moving particles, a scientist's brain may create pictorial representations of moving particles.

A second alleged limitation of scientific theories is, according to Körner (1966, p. 91), due to circumstances that

> . . . a logic which admits only definite individuals and exact classes cannot do justice to empirically continuous series, not even to spatially extended continuous series, the members of which merge into each other either in the sense that they have common neutral components (if they are parts of a complex individual) or in the sense of being connected by having common neutral candidates (if they are classes). In other words, hypothetico-deductive theories, which are embedded in an unmodified two-valued logic, not only replace temporal by atemporal structures, but inexact by exact structures. If continuous gradation and in

[3] In fact, my new brain-model (*cf.* Wassermann, 1974, and in preparation) has the dual capacity of de- and retemporalising engram sequences.

particular continuous change is to be expressed, the exactness of classes and the definiteness of individuals must be supplemented by a postulate of mathematical density, which inevitably increases the distance of the ideal counterparts of empirical individuals and classes from experience.

The preceding passage does not take into account that much of hypothetico-deductive theorising relies, in practice, already on approximations (see pp. 1–2) which involve evaluative decisions. These evaluations cannot be simulated by formal logic, and the use of approximations in deduction procedures does not replace something inexact by something exact. In fact, it does the opposite. Körner's passage seems to refer to the formulation of 'empirical generalisations' (that is, to the allocations of scientific observation to classes). These generalisations only form the *final* deductive statements of hypothetico-deductive theories, as will be explained later. Körner's emphasis on the failure of logical approaches to 'empirical generalisations' (because of the use of exact classes) is a little surprising, since in most applications of theories even their *main* hypothetico-deductive stages cannot on principle be simulated by formal logic (see pp. 1–2 above).

When discussing 'empirical generalisations', Körner's arguments disregard also the facts discussed in § 1.3 (*cf.* Brain's passage, p. 18), which suggest that human brains have, and normally utilise, the capacity to allocate a 'continuous' range of variants to the same class. Without this ability concept formation would not be possible, since a concept *represents* in token form a class of variants. In a cartoon of some celebrated person, usually a few outline strokes are sufficient to allow identification (that is, classification), although the cartoon may differ widely in fine detail from the minutae of the actual face as seen, for instance, in a photograph. Yet the cartoon and photograph are allocated to the same class (by most normal brains).

Any anatomist knows that drawn textbook illustrations of cell types (such as typical neurons or typical osteoblasts) are (like cartoons of men) simplified representations (see p. 63), and that there exists an indefinite range of variation of details. Yet the brains of anatomists can clearly distinguish a neuron from an osteoblast. More generally, sophisticated biologists are rarely interested in definite individuals for their own sake and, apart from idealised textbook illustrations, do not pretend that there are exact classes, and their work never relies on a two-valued logic. Many biological descriptions implicate a class of varying members. Even the Watson–Crick DNA helix (*cf.* Wassermann, 1972) is typical of a wide class of (denumerable varying) structures, each having different (denumerable) base-pair sequences as the structure *per se* does not require *specific* base-pair sequences. Again, the variability of typical electronmicrographs of particular anatomical structures (such as synapses) provides excellent examples (*cf.* Colonnier, 1968, for good illustrations).

If, as all evidence suggests, the capacity to classify is a typical property of human brains, then a deep understanding of this capacity can only be reached *via* neuropsychology, but not *via* philosophy unaided by neuropsychology (see pp. 2–3). No amount of logical reformulation at a purely philosophical level, and divorced from neuropsychology, could help us to understand the manner of attainment of those classifications which are directly carried out by brains, that is, without observing machinery that assists certain methods of classification. Formal logic cannot tell us how brains allocate an indefinite number of variant members to the same class. Only brain-models can provide the required explanations. Formal logic certainly depends on class-representing constructs (that is, concepts) which are *generated* by brains. However, this does not imply that brains form their class-representing constructs by using either formal logic or any type of mathematics. In other words: the existence of a (brain produced) logical 'calculus of classes' does not suggest that brains classify perceived things by using this calculus of classes or any modification of it. It would be like saying that factories which make screws must use screws as raw material in order to make screws.

Some philosophers believe that they can *describe*, by appropriate logic, the relationship between classes and class members. However, even these descriptions, like any classificatory process, are bound to be arbitrary, as will emerge in due course. One purely philosophical approach to the classification problem is due to Körner (see p. 20 above). In this the concept of 'empirical continuity' plays a key role. I shall argue (on p. 50) that this concept is purely metaphysical. It is therefore not surprising that (to cite Körner, 1966, p. 54) 'the analysis of "empirical continuity" has met with apparently insuperable obstacles'.

Körner assumes (see p. 20 above) that there exist 'empirically continuous series' . . . 'the members of which merge into each other', and that their description demands a logical framework which utilises 'mathematical densities' (that is, continuous statistical distribution functions). However, the assumption that there exist 'empirically continuous series' appears to me to be mistaken from the start (see my remarks on p. 50 concerning 'empirical continuity'). One can hardly give a valid logical description of terms like 'empirically continuous series' or 'empirical continuity' which lack adequate explicit definitions (see p. 50).

Philosophers of science who, like Körner (1966, p. 91) and others, try to introduce 'mathematical densities', in order to describe allegedly 'empirically continuous series' of observations (see p. 50 for comments), are not introducing anything basically new. Closer inspection suggests that they try to restate in formal logical terms something that has been common practice in most fields of science for many years. For instance, curve fitting procedures and related ideas (see § 2.4A) allocate (often) *densely distributed* results of repeat observations to a class. The allocation is done on the basis of hypothesised statistical distributions. A curve which is fitted to a set of repeat observations may be regarded as a classifier of these observations (within socially agreed limits). However, curve fitting procedures (and so on) depend on numerous implicit hypotheses, *which introduce an element of arbitrariness* (*cf.* p. 67 and p. 72), that no amount of logical description can remove.

As there exist monographs which deal with the statistical treatment of observations, and with extrapolation from observations (such as curve fittings), it is difficult to see what a logical redescription could add to this enterprise, particularly if it does not bring out its arbitrariness. This arbitrariness is concealed in Körner's passage (p. 20 above), since he does not mention that there exists no unique method of selecting 'mathematical densities' (for example, in curve fittings). One is left with an indefinite number of possible choices, and those made by scientists are convenient, but not logically or empirically necessary. At best such choices seem 'reasonable'. But what is 'reasonable' is in the eye of the beholder. I conclude that if there exists a limitation, it is not the one Körner envisages, but a limitation caused by an arbitrary, though possibly not unreasonable, choice of scientific classification methods.

It is certainly a challenging neuropsychological problem to understand how brains allocate an indefinite number of variants of a stimulus configuration (and the like) to a single class. However, even events which are *dis*continuous (such as the rapid successive representation of different static frames of a film on a screen, or any other 'phi-phenomenon') can be represented in brains so as to *appear* to be continuous. We therefore cannot conclude that physical things, which are *classified* as continuous by some physical systems (such as brains), are continuous *per se* (see also p. 50).

Körner believes that there exists another limitation of hypothetico-deductive scientific theories. He attributes this to Gödel's theorems. The latter are relevant to the consistency of mathematical systems used in making deductions from given scientific hypotheses. The essence of Gödel's theorems is, to cite Körner (1966, p. 93), that

> any sufficiently rich axiomatic system is . . . unable to characterize even the structure of its ideal subject-matter completely, because it is always satisfied by two models which are not isomorphic.

I believe that the threat to scientific theories posed by Gödel's theorems lies more in the realm of very remote possibilities but is extremely unlikely to affect science. In fact, up to now no inconsistencies in any known scientific theory have been traced back to the unwitting simultaneous application of nonisomorphic mathematical structures, based on the same set of axioms.

It should be remembered that if a number of alternative interpretations of a phenomenon exist, then this is almost invariably a consequence of the use of alternative *scientific hypotheses* (as distinct from the axioms of mathematical deduction procedures). The problems of experimental artefact (for example in molecular biology), which often create great ambiguities in the interpretation of results and theories based on the results, far outweigh the hypothetical, but in practice unsubstantiated, dangers which might arise from the implications of Gödel's theorems. The complexities of experimental artefact are illustrated elsewhere in my discussion of membrane structure (see Wassermann, 1972, chapter 3).

As the danger of experimental artefact is always present, and as scientific hypotheses *per se* have only the status of *assumptions which may have to be replaced* in the light of new evidence,[4] scientific theories are not concerned with the discovery of 'objective truth' (*pace* Popper, 1967, p. 12). Instead theories are intellectual edifices which are always subject to limitations or doubts on many different grounds (see, for a typical example, Anon., *Nature*, 1971, **334,** 505). Among reasons for doubts, Gödel's findings have played a negligible (if any) role, particularly in experimental biology, compared with ambiguities presented by possible experimental artefacts. The significance of Gödel's theorem is also overshadowed by doubts arising in certain theories from possible accumulations of computational errors. Even if all mathematical systems could be shown to be internally consistent (a demonstration which, within the systems, is in most cases excluded by the work of Gödel and others; *cf.* Kleene, 1952), then it must still be remembered that the testing of many scientific theories is partly based on computational techniques (such as numerical integrations or numerical differentiations) which use approximations involving errors at each step due to deviations from exact expressions.

If one has to carry out a calculation which involves a large number of variables and, say, evaluates integrals of a high order of multiplicity, then even small errors inherent in approximate numerical techniques could rapidly accumulate. The results could, for that reason alone, be open to much more serious doubt (in many cases) than any raised by Gödel's theorem. Even comparison with experiment provides no guarantee that any 'good agreement' could not be due to a fortuitous cancellation of errors, partly inherent in the theory itself and partly in the approximate computational methods. Hence, unless we have numerous independent confirmations for the inherent plausibility of a theory, numerical agreement with one or two experiments alone would not provide a sufficient ground for strong beliefs

[4] A good example is provided by the required revision in theories of DNA replication of the circular chromosome of the bacterium *E. coli.* For quite a few years it was believed that replication of this circular chromosome proceeds unidirectionally from a particular origin along the chromosome. However, recent genetic mapping experiments (Masters and Broda, 1971) suggest that replication proceeds in both directions from a fixed origin (see also Anon., 1971, *Nature New Biology*, **232,** 129, for further comments).

in a theory. (For the same reason a single numerical disagreement cannot be a valid reason for rejecting a theory, which agrees with many other data.)

The fact is that, despite these diverse potential sources of error and doubt, scientific theories have established a remarkable number of relations and a high degree of order among phenomena, often by combining the result of various theories (for example, the remarkably accurate structure determinations of certain proteins were obtained with the help of experiments and the *theory* of X-ray diffraction). I partly agree with Popper's (1967, p. 12) belief that

> infinitely more important for the scientist than the question of the usefulness of theories is that of their *objective truth*, or their nearness to the truth, and the kind of *understanding* of the world, and of its problems, which they may open up for us.

However, neither Popper nor anyone else has provided any criterion—that is, any decision procedure—which would enable people to say when a theory is 'true' and how near it is to the truth (see p. 32). All we can say is that certain theories explain in terms of a relatively small number of hypotheses a wider range of phenomena than other theories. But this is no safe guide to the validity of these theories in the light of further experiments. The answer is that unless we already knew the objective truth of theories we could never decide that they are objectively true, even if we failed to contradict ('falsify') them for centuries. It is for this reason, I suspect, that some scientists would not share this particular aspect of Popper's philosophy, however much they may go along with him on other issues.

Irrespective of the non-demonstrability of an 'objective truth' status of scientific theories, the order which they have established among phenomena is preferable to chaos despite many persisting doubts (which require constant refinements of techniques and continued reexaminations of existing theories). It is theories (see § 3.4), based on linked networks of tentative hypotheses, that provide at any given time our knowledge of scientific phenomena. The steadily increasing number of well confirmed theoretical relationships, and increasing number of apparently mutually consistent theories which scientists have established, give us confidence that continuation of this enterprise is likely to yield more and gradually improved theories as time goes by.

1.5 Scientific *versus* 'common-sense' language explanations

A. Language-usage analysis

Scientists, who deal with complex physical systems such as brains or individual cells, ask and try to answer difficult questions referring to very intricate systems, and expect highly sophisticated answers. This may be contrasted with the approach of many philosophers to some of their problems. Possibly as a consequence of the rapidly advancing pace of science, which demands expert knowledge, many contemporary philosophers have relied on the decoupling of science and philosophy which I already discussed. One group of philosophers restricted themselves to an analysis of language usage by 'the man in the street'. This is often referred to as 'common-sense' philosophy, since it attempts to analyse the sense *commonly* attributed to linguistic expressions by ordinary (competent) language users. (For typical examples, see Mundle, 1970.)

Undoubtedly *some* philosophical problems are pseudo-problems resulting from faulty *use* of language, and elucidation of correct usage can eliminate these problems. Linguistic philosophers, it is argued, are not concerned with improving or altering living languages. They should accept these as they find them, but expose the normal usage of their expressions. Linguistic expressions are only valid when used in appropriate contexts. Some philosophical pseudo-problems originated when expressions which are normally used in familiar contexts were transferred to new contexts where their use was unfamiliar. The word 'power' in the statement 'Napoleon was a powerful man', or 'this is a power station', is clear from common-sense usage. But in the statement 'God is all-powerful', the word 'power' is used in a 'metaphorical' context, differing from contexts in which it is used in everyday situations. We cannot deny that its new usage may convey some metaphorical or poetic meaning to people, and its employment by preachers might suggest, according to the usage criterion, that it *has* a place in the language of 'the man in the street'. But this very example, like many others, illustrates that usage of certain utterances by 'the man in the street' on certain occasions or in certain contexts does not imply a specifiable significance.

Wittgenstein (1953) and others realised that an analysis of the *usage* of language requires no hypotheses. Mastery of the nuances of linguistic expressions and an aptitude for analysis are the only prerequisites. Yet where does it lead to? Granted that the analysis of language could uncover linguistic abuses, could it solve any long-standing problems? One is reminded of Lashley's (1951, p. 230) remarks in another context:

> I attended the dedication, three weeks ago, of a bridge at Dyea, Alaska. The road to the bridge for nine miles was blasted along a series of cliffs. It led to a magnificent steel bridge, permanent and apparently indestructible. After the dedication ceremonies I walked across the bridge and was confronted with an impenetrable forest of shrubs and underbrush, through which only a couple of trails of bears led to indeterminate places . . . .

Paraphrasing Lashley's subsequent passage (which refers to another context), I can only say that I am at a loss to see where further development of Wittgenstein's approach could lead. Many epistemological problems require scientific (and particularly neuropsychological) *hypotheses* (see pp. 3–5 and 6–7) and cannot be resolved by a hypothesis-free analysis of language-usage. However, as ordinary languages contain their own (hidden) hypotheses (see below), language-usage analysis may at least serve to exhibit these hypotheses where they exist.

B. *Black box approaches*

Philosophical language-usage analysis treats man as a black box with information input channels and behavioural output channels. Proponents of this point of view argue that processes which go on 'under the skin' are none of their concern. Black box linguistic analysis corresponds to the zero order approximation of the neuropsychology–philosophy feedback cycle (see p. 3). It therefore provides no answers to problems which require higher order approximations.

The weakness of black box theories can be illuminated by many examples. If, for instance, man and octopus are regarded as two different types of black box, this throws no light on their totally differing capacities for information processing. These can only be elucidated by explaining the differing properties of their nervous systems. The inherent ambiguities of black box approaches, in contrast to detailed mechanistic explanations (for example, in terms of molecular and macromolecular structures resident inside the black boxes), has recently been emphasised in a different context (Anon., 1971, *Nature*, **234,** 380), and other dangers of black box theorising were lucidly discussed by Bunge (1964).

C. *Language-usage analysis in science*

Wittgenstein (1953) and some of his followers considered the correct usage of a linguistic expression in its appropriate context as being analogous to a move in a game, which conforms to the *rules* of the game (Pole, 1958; Sellars, 1954). In fact, Wittgenstein (1953) illuminated language-usage analysis by the construction of artificial simplified language models, which he called 'language games'. Such games illustrate but do not state the rules. Context rules are neither scientific hypotheses nor rules of grammar, but are concerned with stating or exemplifying 'common-sense' usage of language. Grammatically correctly constructed sentences may be nonsensical ('houses lay eggs' or 'the motor car walked slowly down the road' or 'my grandmother is the nicest man I know'). Also, linguistic expressions need not refer to 'real entities' and hence do not entail existence of such entities. If a fairy story states that 'a dragon sat down in Trafalgar Square' this does not imply the existence of dragons. Linguistic analysis is 'behaviouristic' in that it is only concerned with the behaviour of the language perceiver and language producer. Being 'understood' by 'the man in the street' who shares your vocabulary (and so on) is the criterion of correct language usage (*cf.* Midgley, 1955, p. 190).

Language usage analysts may include in their analyses 'artificial languages', such as various forms of mathematics, including formal logic (Wittgenstein called these the 'suburbs' of languages). Parts of the philosophy of science are therefore explicitly or implicitly concerned with language usage analysis of scientific languages. In fact some paradoxes of science could, and probably do, originate in an unjustified use of scientific expressions. Sometimes, for example, scientists state that their work contradicts a specific phenomenon *x* claimed to have been discovered by scientist Y. Yet on closer inspection it may turn out that the *context* (that is, the experimental procedure) within which Y claimed phenomenon *x* to exist, differs completely from the procedures used by those who disagree with him. Hence in such cases it remains doubtful whether the opponents are discussing the same phenomenon. Some scientific controversies centre precisely on such issues.

D. *Hypotheses and scientific languages*

As scientific theories may involve (i) new scientific hypotheses and (ii) the use of ordinary as well as artificial languages (mathematics, and so on), a contradiction within a theory could be due to faulty language usage or to faulty scientific hypotheses or to a mixture of both. Faulty language-usage in science may arise from referring the same phenomenon to different experimental procedures (see above). Again, contradictions could arise from the addition of simplifying assumptions (for example, in order to make certain mathematical equations of a theory soluble). Such simplifying assumptions amount to additional *ad hoc* hypotheses. Because of this complex intermingling of explicit hypotheses, implicit *ad hoc* hypotheses and possible faulty language-usage (see above), a simple language-usage analysis in science, of the type used by 'pure philosophers', is not practicable, and scientific theorising relies on a large number of different procedures to decide issues, particular in cases where disagreements arise. Scientists may reexamine their data or experimental techniques, attempt to obtain independent confirmation of hypotheses by various means, or they may try to alter their mathematical methods of approximation (where mathematics is used) and so forth.

E. *Infiltrations of ancestral prejudices*

Some philosophers of science have raised issues which relate to science as well as to the language of 'the man in the street'. The language of 'the man in the street' contains many expressions which lack precise empirical definitions within experimental psychology or other sciences, although 'ordinary' language terms are often used as if they refer to *causal factors* of behaviour. This has prompted some philosophers of science to raise problems, which, I believe, are not justified. Thus, Feigl (1958, pp. 388–9) in his celebrated essay on the 'mind–body problem', argued that

> to maintain that planning, deliberation, preference, choice, volition, pleasure, pain, displeasure, love, hatred, attention, vigilance, enthusiasm, grief, indignation, expectations, remembrances, hopes, wishes, etc. are not among the causal factors which determine human behavior, is to fly in the face of the commonest evidence, or else to deviate in a strange and unjustifiable way from the ordinary use of language. The task is neither to repudiate these obvious facts, nor to rule out this manner of describing them. The task is rather to analyze the logical status of this sort of description in its relation to behavioral and/or neurophysiological descriptions . . . .

The *fact* that the language of 'the man in the street' asserts that feelings of joy, grief, and so on, are *causal factors* which determine human behaviour is undeniable. It is presumably rooted in interactionist beliefs (in an immaterial mind acting on a material body) of our primitive ancestors, which they incorporated into their languages and which, in modified forms, were passed down to us. We may, without danger, continue to use, for *everyday* purposes, primitive ancestral modes of language, which assign causal status to feelings. But this usage does not justify (a) the assumption that (private) feelings *per se* have causal status (that is, it is no ground for believing in interactionism); (b) efforts to relate the linguistically embedded implicit hypotheses of our unenlightened ancestors to behavioural or neurophysiological or other biological concepts; and (c) attempts to uphold the implied interactionist hypotheses of the language of 'the man in the street' by logical manoeuvres (compare my discussions of Scriven's arguments in § 4.7B). Feigl's request, to analyse the logical status of the language of 'the man in the street' in relation to neuroscientific concepts, would only be justified if the language of 'the man in the street' contained no implicit causal hypotheses of its own (for example, that 'desires' or 'greed' *cause* people to act). I think that there are no glowing prospects of relating primitive ancestral modes of thinking (which are retained in current language) to sophisticated concepts of neuropsychology. It might be like trying to relate the beliefs of witch doctors to modern medicine, or attempts to validate the delusions of schizophrenics.

A real danger arises elsewhere. It is at present, and possibly indefinitely, essential and convenient to use parts of the language of 'the man in the street' in conjunction with scientific languages in the description and evaluation of experiments. In neuropsychology this harbours dangers which do not exist as obviously, if at all, in physics or molecular biology, chemistry, and so on, since ordinary language is crowded with hidden primitive hypotheses (see above) about 'mental entities'. (Fortunately our primitive ancestors made no implicit statements about the causal behaviour of quasars or ribosomes, so that such statements have not entered our everyday language.)

F. Different levels of explanation

My advocacy of a successive approximation approach based on a neuropsychology–philosophy feedback cycle will, if adopted, require successively deeper levels of explanations involving increasingly sophisticated hypotheses. The gulf which at present separates scientists from philosophers of language-usage (who rely on decoupling of science and philosophy) was brilliantly illuminated by the late C. A. Mace in an editorial foreword to a book by M. D. Vernon (1962). He presented as an example the striking differences of approach between 'common-sense' philosophers and scientists to the problems of perception. He noted that:

> To 'the man in the street' it may seem idle to discuss the manner in which he perceives the world around him. It requires study to see the problems with which the psychologist is concerned when he sets out to explain why we see things as we do. This 'man in the street' has been encouraged in his doubts about the need for explanation of the facts of perception by some sophisticated philosophers. These philosophers take the line that the only facts of perception which call for explanation are the abnormal facts such as illusions. They say that we can properly ask for an explanation of the fact that two lines which are in fact equal appear different in length, but we do not need to ask for an explanation of the fact that two lines that are equal appear to be equal. So too they say, we can properly ask why snow which is white looks yellow, but we cannot properly ask why snow which is white looks white. They say that when two lines that *are* equal appear to be equal it is a sufficient 'explanation', if this is to be called 'explanation', to say simply, because they are equal. So too, *mutatis mutandis*, with the white snow.
>
> The difference between the psychologist on the one hand and the man in the street and these philosophers on the other arises perhaps from the failure of the latter to do justice to the fact that there are many different *kinds* of explanation. There *is* the kind of explanation of the fact that grass looks green—which may satisfy the philosophers and the man in the street—which is simply that grass looks green because it *is* green. There *is* the sort of explanation which appears to satisfy some chemists and some botanists which takes the form of saying that grass is green and looks green because it contains chlorophyll. There is another kind of explanation which satisfies physiologically-minded psychologists which is given in terms of rather complicated 'theories of colour vision'. Perhaps also some notice should be taken of the kind of explanation which satisfies children and devout old ladies who would say that grass is green and looks green because God arranges things so because green is a restful colour.

1.6 Linguistic idiosyncrasies *versus* scientific generality

Perhaps one feature more than any other distinguishes scientific theories from 'common-sense' language analysis. In the language of 'the man in the street' a particular linguistic expression can often have many different shades of meaning when used in different contexts. Correct usage of an expression can therefore only be ascertained by listing piecemeal all known contexts in which the expression is commonly used by people. Gellner (1959) called this many-sidedness of linguistic expressions 'polymorphism'. Because of this, each specific usage of an expression can often only be ascertained for a specific range of contexts and, as a rule (but see Masterman, 1957) no *wide* generalisations about use of expressions are possible in common language. In the language of 'the man in the street' nothing can be discarded as being unimportant, and the particular and idiosyncratic use of each word may count.

Scientific theories, on the other hand, always *generalise* and in the process of generalisation may also *idealise*, by disregarding empirical features which the theorist may, rightly or wrongly, consider as being of 'minor significance', either for the time being or for a theory which intends only to correlate major aspects of certain phenomena. Idealisations allow us to encompass many variants of phenomena within a single, often reasonably well definable, *class*, simply by concentrating on *general* class-characteristics and by disregarding idiosyncratic features which are only characteristic of a few members of a large class. In this way scientists often separate the significant from the trivial, although some features which at one time are deemed to be trivial may, in the light of later investigations, turn out to be more significant for the whole class of phenomena than was thought at first to be the case.

Linguistic philosophy, as applied to the language of 'the man in the street', treats (as a rule) everything in a language as being equally important. In contrast to this, scientific theories, apart from introducing hypotheses, deal with class-characteristics of classes of phenomena and systems, and with relations between classes of phenomena. As a rule, scientists are not concerned with the unique and idiosyncratic. They are not interested in the genetic propagation of a disease through the family tree of a *particular* human family, for its own sake, but to see how a particular disease propagates through the family trees of many families (in order to establish the dominance *or* recessiveness of certain genes, *cf.* Becker, 1964). Scientists are not concerned with listing and observing the properties of every star in our galaxy (even if this were possible), but with classifying and explaining the properties of detectably different *classes* of stars (such as white dwarfs or pulsars). Likewise, zoologists are primarily concerned with explaining and exploring those properties of animals which are common to a wide class, but not with the properties of a particular species, unless similar properties are likely to be shared by many species. Thus one is not interested in discovering the precise DNA sequences of the chromosomal DNA of every butterfly species (even if this were possible) but rather in general principles which enable us to understand how a butterfly's DNA could generate developmental patterns, and one wants to know whether similar principles apply to mice and men. Again, pure psychologists are not interested in studying particular habits of individual people for their own sake. They wish to find out how people and animals learn, and which mechanisms are operative in learning processes, and how these mechanisms differ for different species. However, what a particular man learns during his lifetime (while possibly of interest to psychiatrists and their assistant applied psychologists who may wish to decondition him) is of no interest to a pure scientist *per se*.

1.7 Knowledge as a system of hypotheses: Popper's epistemology and its limitations

My point of view concerning a theory of knowledge agrees in parts closely with that of Popper. Wisdom (1964, p. 123) has expressed Popper's outlook as follows:

> Popper's theory of knowledge, as is widely known, is very simply that all we ever have at the very best is a set of hypotheses; however well authenticated, they are never more than that. The best-established piece of knowledge may have to be modified or even given up. We may retain it only so long as our attempts to test it have failed to refute it. [Popper (e.g. 1970, p. 2) considers 'knowledge' as 'public knowledge': for example, available in libraries.]

At first sight the quotation from Wisdom may appear difficult to reconcile with Popper's own statement (cited on p. 25), that scientific theories are concerned with *objective truths*. However, Popper (who, like Bunge (*cf.* 1967, p. 1–6) and others, including myself, has a personal metaphysical belief in an autonomous external world), while accepting the view that we may frequently have to replace existing parts (such as subhypotheses) of theories, or even complete theories,[5] believes, I presume, that nevertheless the process of successive 'improvement' converges towards an 'ultimate' limit in theorising, which Popper considers to be the *objective truth*. This, at least, is my interpretation of Popper's philosophy. From this view I differ in one important respect (see p. 25). I concede the *possibility* that science is 'progressive', in the sense that the succession of theories may well tend towards a 'limit' (not in the mathematical sense, but in the sense of reaching a stage where 'improvements' in theories may become very slight). But I believe that we can never know whether such a limit exists, and if it exists how near we are to it (see also p. 25).

Now for some further comments on Wisdom's (above) and Bohm's (see note 5) views concerning Popper's theory of knowledge. Popper has rightly emphasised (*cf.* Popper, 1963a) that 'falsifiability' of a hypothesis or theory, by newly discovered contradictory evidence, forms one yardstick for the constant reassessment of existing hypotheses and theories.[6] But there lurks a danger of serious misunderstanding. For example, I once heard a distinguished scientist start a lecture with the words 'I have for the last twenty-five years tried to destroy (falsify) the theory of Professor X'. In a similar spirit I was once asked, after establishing a particular theory, 'how can one

[5] Bohm (1964, p. 216) observed that: 'The need for the falsifiability of theories, as emphasized by Professor Popper, already shows that the question of trying to determine the meaning of truth cannot properly be treated by the direct approach. Thus, if every really acceptable theory must be falsifiable, then it seems almost certain that, in time, such a theory will *actually be falsified* as more accurate experiments are done in broader domains and in new contexts. (As has in fact, already happened to an extremely large number of theories in every branch of science.) . . .'

[6] 'Falsification' of hypotheses does not (as a rule) proceed directly *via* formal logic, but depends on significance tests or other agreed criteria.

destroy your theory?' Such peculiar attitudes stem, I think from serious misunderstandings of Popper's point of view. They show the inherent dangers of Popper's arguments when used by others who are neither professional theorists themselves nor well versed in the philosophy of science and who have misconstrued the spirit of Popper's philosophy. I suspect that most scientific theorists do not primarily invent new hypotheses and theories in order to have them *destroyed*. In fact, if this were the true motivation of scientific theorising and observation, then it would be hard to reconcile with Popper's credo that through successive criticisms we can approach the 'objective truth'. (See also Achinstein, 1968.)

Just as composers, novelists and playwrights do not produce their works essentially for destructive attacks by critics, but for the enjoyment and benefit of mankind, so scientific theories are primarily intended to give us a deeper understanding (in a sense to be explained later) and, if possible, to help us in the practical prediction of phenomena, thereby aiding applied science. I believe it is correct to assume that scientific theories are mainly intended as *constructive* efforts. I presume that Wagner did not compose *Tristan und Isolde* and *Der Ring des Nibelungen* for the benefit of vitriolic critics like Eduard Hanslick (immortalised in the guise of Beckmesser in *Die Meistersinger von Nürnberg*) but for the enjoyment of the art-loving public. I also believe that Dirac did not invent quantum electrodynamics and the theory of the spinning electron primarily for the purpose of having these theories destroyed. Again, as far as I can gather, Crick and Watson did not receive a Nobel prize mainly because their DNA structure had provided something that could be falsified.

The misunderstandings of those who attach excessive weight to Popper's falsification criterion originate probably from disregarding two essential features of scientific theories.

(1) Many hypotheses of science appear to survive indefinitely, at least in *essentials*, although their *detailed forms* may become gradually improved. The misguided idea that hypotheses or theories are dealt with in all-or-none fashion simply does not apply to many types of hypotheses or theories. Because many theories idealise, improvements often take place by replacing existing hypotheses by more 'realistic' ones which take more of the observed features of phenomena into account. (For instance, for many purposes—such as in chemical calculations—Schrödinger's equation forms a perfectly adequate hypothesis, provided one builds spin into the Hamiltonian by means of systems of Pauli matrices, instead of using the fully relativistic Dirac theory. Hence the 'falsification' of Schrödinger's theory by the evidence which led to Dirac's equation does not mean *either* that Schrödinger's theory has lost its uses or approximate validity (within certain contexts) *or* that it was not an important milestone on the way to a better theory.)

(2) The plausibility of a hypothesis or theory may not only be falsified but

it may also be *positively reinforced* by new confirmatory evidence. It is a totally misleading view that most scientific experimentation is done in order to falsify existing theories. Much (or possibly most) experimental work is done in order to obtain further confirmation of existing hypotheses. It sometimes happens in the process of searching for confirmatory evidence that apparent 'falsifications' of existing theories occur. However, even then care is required, since it may turn out that the supposed falsification does not genuinely dispense with the *essence* of an existing hypothesis or theory, but that it may require only minor modifications of these hypotheses or theories.

Let me also emphasise that the problem of confirmation or refutation of hypotheses of *complex and comprehensive* theories is not as straightforward a problem as some philosophers of science envisage (see below). Some theories (such as non-relativistic quantum mechanics) are so strongly empirically anchored, through so many experiments, that the problem of 'falsification' of any part of the theory may require as much critical appraisal *of the criticism* as the proposition which it criticises. Comprehensive theories (such as the theory of evolution) are more often than not gradually modified rather than totally replaced, and the falsification criterion is sometimes repeatedly applied to various subhypotheses of the theory, but not to the theory as a whole, which in total may in its essentials be increasingly confirmed. I conclude that confirmations are as important as falsifications of parts of a scientific theory.

1.8 Different levels of knowledge

Whether one wishes to subscribe to Wisdom's (1964) verdict, that Popper has overcome the Kantian 'problem' of epistemology, must remain a matter of personal assessment. Wisdom remarked that the 'problem is solved [by Popper] by being shown not to arise in the form in which it presented itself to Kant'. There remains the possibility that, while Kant may have posed a problem which Popper and other philosophers may regard as wrongly posed, this provides no assurance that the problem of epistomology as seen by Popper (beyond science) supplies a sufficiently wide basis for epistemology.

I like to look upon epistemology in a way resembling Mace's way of looking upon the various views of perception (see p. 30). I suggest that we can pose and answer epistemological questions at different *levels*, and that we have to examine the *decision procedures* used for assessing 'answers' according to the level of enquiry concerned. Thus, different 'answers' to the same epistemological question may *satisfy* different kinds of enquirers (Neanderthal man and a modern biochemist would, I suspect, not be equally satisfied with answers given by the former relating to the origin of certain diseases). The view just expressed might be called the hedonistic approach to epistemology—that is, at different epistemological levels one poses different *kinds* of hypotheses and adopts different decision-making criteria for judging answers given, the type of criteria adopted depending on the satisfaction which they give to their user. (I am here concerned not with the pragmatic value of hypotheses but purely with decision procedures adopted for evaluating their acceptability, however useless such hypotheses might be for the practical affairs of the world.)

For a scientist, 'knowledge' often implies a network of hypotheses of a theory which relates *classes* of phenomena (see below). Some philosophers, as well as scientists, believe that 'direct knowledge' ('private experiences') should also be included in the discourse of scientific knowledge. For reasons stated in § 2.1, I am not optimistic that such efforts are likely to be crowned with success.

It is a plausible scientific hypothesis that human brains have the capacity to operate as hypothesis-formulating automata, since brains are (in the opinions of non-interactionists) capable of creative thinking, of which hypothesis-formulation represents a particular aspect. But we must appreciate that human brains can formulate 'empirically anchored' as well as metaphysical hypotheses. Some of our everyday hypotheses (*cf.* Peters, 1958) relate to the idiosyncratic behaviour of other people. Such hypotheses are, more often than not, only feebly empirically anchored, and hence neither adequately confirmable nor falsifiable. (Asking people to account for their behaviour may not elicit a truthful answer.) Hypotheses about *individual* human behaviour are (except in the case of psychotics) often no more than metaphysical assertions, although they may be very satisfying to their users.

Again, historians have put forward hypotheses concerning the activities, 'motives' and relationships of various long-deceased individuals (on the basis of historical 'evidence'). But even an autobiography may be regarded as being partly hypothetical (owing to the possibility of deliberate misrepresentations or memory failure). In cases where indirect corroboration is ruled out, some autobiographical accounts may be on a par with statements about fictitious entities (dragons, and the like). At another level we have religious beliefs, which are *metaphysical hypotheses* (such as divine omnipotence). Systematic procedures for evaluating answers to questions may exist at one level of epistemology (for example, in science), but be lacking and replaced by authoritarian dogma or personal belief at other levels of epistemological enquiry.

Knowledge acquisition and evaluation involves (i) environmentally supplied information, and (ii) brain mechanisms of individuals. Brains can form and allocate values to hierarchies of concepts which are either self-generated or related to the environment. Which particular hypotheses and evaluative criteria are considered as being 'satisfying' depends on the brains of individuals, on their education (that is, the types of memory traces and critical evaluative practice procedures which their brains have formed) and on the problems they are confronted with. Even scientists who may agree on the *general* modes of decision procedures adopted in science, may violently disagree with the interpretations of some experiments and their bearing on hypotheses. Some people say that certain metaphysical hypotheses are satisfying to them while the same hypotheses are repugnant to others, because they prefer to restrict themselves to a different *level* of epistemology.

If all knowledge ultimately depends on hypotheses which in the sense of formal logic can never be proved to be true, then there is no reason why somebody could not simultaneously accept various hypotheses at different, non-overlapping, epistemological levels. He could uphold metaphysical (or partly metaphysical) hypotheses about the 'motivations' of his uncle's behaviour, empirically falsifiable or confirmable hypotheses about elementary particles, purely metaphysical hypotheses concerning religious statements, provided that he adopts decision criteria (or lack of these) appropriate to the class of hypotheses concerned.

Suppose that a scientist accepts (privately) divine omnipotence. If you ask him to explain the processes which cause innocent children to die of virus diseases, he could try to explain these in terms of scientific hypothesis. If you ask him why God, who is omnipotent, let these children die, or 'where was God at Auschwitz?', he may say that God's ways are unknown to him. In other words if you ask a metaphysical question he could plead ignorance, but he could not give you a scientific answer. After all, scientists also plead ignorance in innumerable scientific matters. One could not expect to 'explain' empirical evidence (such as evidence for the existence of large numbers of genocidal Germans at one time of history) in terms of metaphysical hypotheses (such as the omnipotence or omniscience of God). More generally, the question of the existence of evil in the world, in the face of religiously pronounced divine omnipotence, is unanswerable. Human usage of the word 'evil' refers to socially or individually accepted moral choices (made by one or more brains)—that is, it refers indirectly to decision processes by empirical systems. However, empirical matters relate to systems of hypotheses, whose status differs from those of metaphysical hypotheses.

Particular levels of epistemological discourse are more appreciated by some than by others. We cannot reproach the lady who believes in the metaphysical hypothesis 'that grass is green and looks green because God arranges things so because green is a restful colour' (p. 30). Such metaphysical hypotheses are based on purely personal decisions. Which type of hypothetical system one prefers may depends on the requirements of the moment and on idiosyncrasies of brain composition. A criminal may adopt different hypotheses concerning the morality of his behaviour from a judge and jury. At a tea party, where one might gossip about people, a different level of hypothesising about the behaviour of human beings is apt to be adopted from that at a conference on experimental psychology or at a theological conference. It is depressing, but I believe true, that we cannot detect any absolute criteria for deciding answers to questions (typically for deciding the validity of hypotheses). Otherwise possibly some political feuds, religious wars, genocide and many other results of controversies might have been eliminated. But even this seems doubtful, since human brains can produce agressive behaviour, as well as creative thoughts directed towards scientific understanding of agression. (It may be that the tendency of brains to produce agression would lead to human destruction even if philosophical agreement on numerous issues could be reached.)

The significance of the existence of different types of hypotheses at different epistemological levels appears also in another context. Sometimes people wish to *know* the 'cause' of some particular events in somebody's life. Here one could build up an ever regressing sequence of diverging causal chains. In contrast to this, scientific knowledge, being concerned with classes of scientific phenomena, is not, without *ad hoc* hypotheses, suited for *applying* theories to individual cases (see § 3.6A). Even if one wishes to *apply*, say, electromagnetic theory to wave guides, one must make *ad hoc* assumptions about prevailing boundary and initial conditions of the system. The more complex a particular system, the more difficult it is, without ever mounting *ad hoc* hypotheses, to *apply* general class statements of science to the interpretation of the behaviour of individual members of a class (*cf.* my criticisms of the *applications* of psychoanalytic theory in § 4.2). In fact, my arguments on p. 29 concerning Feigl's suggestion that we should analyse 'common-sense' language usage in relation to neuropsychological descriptions are also centred on the difficulties of comparisons of hypotheses associated with different epistemological levels. Everyday language, as mentioned, has numerous hypotheses (of a low epistemological level) built into it, which one cannot compare with the high level hypotheses of (say) neuropsychology.

1.9 Rationality, criticism and brains

Bartley (1964) provided an interesting discussion of various attempts to 'justify' rationality against claims of its logical limitations by philosophical sceptics and fideists. The latter argue (to quote Bartley, 1964, p. 5) that

> *from a rational point of view, the choice between competing beliefs and positions and ways of life, whether scientific, mathematical, moral, religous, metaphysical, political, or other, is arbitrary*. In short, it is claimed to be demonstrable by rational argument that it is logically impossible to act and decide on rational grounds when it comes to such choices—even though the making of such choices in a nonarbitrary way can be considered to be the main task of rationality.

Bartley (1964, p. 5–6) states further that the argument used by sceptics and fideists hinges on an analysis 'of what is commonly regarded as the rational way to defend ideas', and that it takes the form of challenging the rationalist with a non-terminating regression of questions of the type 'How do you know?', 'Give me a reason', or even 'Prove it!'

After criticizing Ayer's (1956) attempted defence of rationalism, Bartley (1964, p. 20) fell back on Popper (1960), by asserting that:

> The traditional questions of philosophy are authoritarian in structure in the sense *that they all beg authoritarian answers*. Questions like 'How do you know?' 'How do you justify your beliefs?' or 'With what do you guarantee your opinions?' demand authoritarian answers, whether those authorities in particular cases be the Bible, the leader, the social class, the nation, the fortune teller, the Word of God, the intellectual intuition, or sense experience [presumably in the form of 'private experience']. A demand for a justification or a guarantee cannot be answered except by providing something authoritative in the sense that it is unquestionable, does not itself need justification, and hence can guarantee the correctness of a conclusion.

In contrast to this, Popper urged '*criticisms* of standard' as the main task of the philosopher. 'Philosophers, he argued, should not demand and search for infallible intellectual authorities, but should instead try to build a philosophical program for counteracting intellectual error.' (Quoted from Bartley, 1964, p. 21).

However, if one accepts 'nonjustificational criticisms' as suggested by Popper and elaborated by Bartley (1964, pp. 29–31), this amounts once more to accepting an authority of a different type, namely acceptance (or rejection as the case may be) of the views of a *critic*. Criticisms sometimes turned out to have been totally unjustified. Yet Bartley (1964, p. 21) asks:

> How can our intellectual life and institutions be arranged so as to expose our beliefs, conjectures, policies, sources of ideas, traditional practices, and the like—whether justifiable or not—to maximum criticism, in order to counteract and eliminate as much intellectual error as possible?

I believe that this type of argument only shifts scepticism to a new position. We are confronted with the additional need for a decision procedure which assures us whether a critic is right or wrong. The sceptic will ask: what is the critic's decision procedure and what justifies it? Criticisms, after all, are presumably carried out by human brains, and the fallibility of brains as thinking machines is well known. Many criticisms of theories or discoveries turned out to be inappropriate and to have retarded the progress of science. The new anatomical descriptions of Vesalius were the subject of derision in his day. People who attacked his findings based their criticisms on the validity of Galen's writings (that is, on authoritative standards; *cf.* Haggard, 1934). Let us turn to another one among numerous examples. Although a wave theory of light was advocated by Hooke, and subsequently, in a superior version, by Huygens (*cf.* Born and Wolf, 1959, p. xx), 'the rejection of the wave theory [of light] on the authority of Newton led to its abeyance for nearly a century' (cited from Born and Wolf, 1959, p. xxi). The critique of Newton's followers was founded on the corpuscular theory of light, which explained certain optical phenomena in terms of the straight line motion of light particles. It was not until Fresnel's work appeared (see Whittaker, 1951, p. 108) that the wave theory of light was accepted. (For one of its best modern expositions, see Born and Wolf, *Principles of Optics*, 3rd edition).

Again, Tyrrell (1947, p. 229) states that the medical profession of about a century ago refused to admit the genuineness of hypnosis:

> When the most painful surgical operations were successfully performed in the hypnotic state, they said that the patients were bribed to sham insensibility; and that it was because they were hardened imposters that they let their legs be cut off and large tumours be cut out without showing any signs of discomfort . . . .

It would be totally wrong to think that it was the church or any other theological authority who pronounced such criticisms, but it was the opinion of scientists themselves who in these, and many other cases (such as the views raised against Jenner's discovery of vaccination), put forward misdirected criticisms. To assume that criticism is ever unbiased is logically as unjustified as any unshakable belief in authority. To assume that criticism must not itself be justified is tantamount to admitting *ad hoc* scepticism, since anyone who does not like any theory or fact, could simply say 'I don't believe it', without having justified his views. (Such *ad hoc* scepticism is in fact not uncommon in private scientific discussions, and appears even at scientific conferences in thinly veiled form.)

Moreover, even if one day we should understand adequately in terms of theories of brain function how brains reach decisions, how they evaluate theories and statements, such theories could only hypothesise *how* brains can reach decisions (for example, establish criticisms). But they cannot assure us whether any decisions or evaluations made by brains are valid. Criticisms of standards themselves demand standards relative to which the criticisms can be exercised, and standards are arbitrary (that is, authoritarian). A 'non-justificational criticism' amounts to a 'standard free' criticism, and nobody has explained how such criticism could operate in any but *ad hoc* fashion.

In any case, Popper (*cf.* 1970) is patently aware that the question and answer game (that is, an answer to one question often raising a new question) is not a particular privilege of philosophical sceptics, but forms also a fundamental aspect of scientific research. Criticism of a scientific theory—that is, dissatisfaction with some of the answers given by a particular theory—and requests for further explanations of some of its aspects, are really a form of scepticism. However, the difference between scientific and philosophical scepticism is very marked in one respect. Some philosophical sceptics would like to attain 'absolute' certainty, while the scientific sceptic only asks for better, and to him more satisfying, answers, without demanding any certainty of their nature. His standards of criticisms are taken relative to (a) empirical data, and (b) the general methodology of theory construction, which both conform to quite generalised (though partly arbitrary) procedures.

I believe that the acceptance of 'standards' for evaluation in science does not mean that we are returning to the middle ages, although the real danger in science, as elsewhere, is dogmatism—that is, the tenacious adherence of people to one set of hypotheses, to the extent that they will not be prepared to consider alternatives. Much of this, unfortunately, is connected with the personal standing of scientists.

If a celebrated scientist puts forward a weakly supported hypothesis, it is likely to receive more attention than a strongly supported counter-hypothesis advanced by a little known man. The contemptuous treatment which Mendel received at the hands of some of his now almost forgotten, but then well known, contemporaries provides a good example that 'criticism' as envisaged by Popper, while excellent in principle, is liable to be biased, because human brains are evaluative systems which have built-in biases (or prejudices). Brains can produce scientific, racial or many other kinds of prejudices (*cf.* Allport, 1954), and this makes the 'eradication of intellectual error' by 'criticism' sometimes a slow enterprise, since many people are only prepared to believe what they wish to believe.[7]

[7] For instance, Philbrick and Holmyard (1945, p. 5) wrote: 'However laudable unquestioning submission to authority may be in the disciple of a revealed religion or in the citizen of a state, it is a serious blemish in the man of science. Dalton's anathema on Gay-Lussac and Avogadro delayed the progress of chemistry by nearly half a century.'

Philbrick and Holmyard (1945, p. 51) tell us: 'How the atomic theory was to be reconciled with Gay-Lussac's law could not be perceived by Dalton, simply and solely because the idea of the indivisibility of the atom had become an obsession with him. Thus his biographer Henry tells us that, to clinch a certain argument, he remarked with an air of finality: "Thou knows *it must be so*, for no man can *split an atom*." . . .

'Avogadro's supreme contribution to chemistry was his suggestion that a distinction might reasonably be made between the ultimate *chemical* particle of an element, the "atom", and the ultimate *physical* particle of a substance, the "molecule". While accepting the Daltonian indivisibility of the atom, he was able to visualize molecules of atomic dimensions which might or might not be divisible according as they consisted of more than one atom or of a single atom.'

Again, Philbrick and Holmyard (1945, p. 52) noted that: 'Unhappily, Avogadro's brilliant suggestion was neglected for nearly fifty years, partly owing to the fact that he published it in a somewhat obscure journal, but mainly because Dalton's authority in theoretical chemistry was supreme and unchallenged.' (For the further history of the subject the reader is referred to Philbrick and Holmyard, 1945, p. 53.)

2 EMPIRICAL STATEMENTS AND EMPIRICAL GENERALISATIONS

2.1 The distinction between 'private' and 'public' worlds

A. Preliminaries concerning 'public' observations

In what follows I shall maintain a sharp *classificatory* distinction between (a) the world of 'private experiences' (or 'subjective observations' as some people call them—*cf.* Ryle, 1949, p. 222; Schlick, 1935; Weinberg, 1936, for examples and discussions), and (b) the world of things or events which are accessible to 'public' (or 'intersubjective') observations. Scientists are, at the observational level, exclusively concerned with 'observing' things which are in principle accessible to those who are equipped with appropriate instruments, and who have acquired competence in the techniques needed in using this equipment for particular types of observations (*cf.* Mercier, 1970). In some cases the 'observing equipment' may be simply the scientist's sensory surfaces and his brain (for example, when a psychiatrist or clinician gives a report on a patient's behaviour). In other cases a sophisticated laboratory or a costly astronomical observatory or other large installations (such as a radio telescope or elementary particle accelerators) are involved which contain the observing instruments (including electronic computers).

Many other impressive examples spring to mind. One of these is provided by the complex techniques and machinery devised by R. W. Sperry and his associates for preparing and studying 'split brain' animals (*cf.* Gazzaniga, 1969). The techniques used by Eccles (*cf.* 1964a) and others for studying synaptic transmission, Hodgkin's (*cf.* 1964) and others' methods for observing nerve impulse transmission in single 'giant' neurons, provide further instances, chosen almost at random from among a long list of twentieth-century laboratory techniques. In other cases 'public observations' may be performed in the natural habitat of living organisms. This applies to the work of ethologists (*cf.* Hinde, 1966).

A few examples may help to illustrate that scientific 'observation' is, as Mercier stressed, in reality confined to a limited circle of workers conversant with the required, often highly sophisticated, technical skills. Use of the Kleinschmidt *et al.* (1962) technique for making wholemount preparations of chromosomes (*cf.* Du Praw, 1965 for typical applications), employment of column chromatography, the techniques used for determining amino acid sequences (*cf.* Sanger and Tuppy, 1951a, 1951b), or nuclear magnetic resonance studies in molecular biology, provide typical examples of one kind. Detection of specific ionisation tracks in photographic emulsions (attributed to 'hypothetical systems' such as mesons and other postulated elementary particles), low-temperature physics, the use of X-ray telescopes, lasers, and a host of other machinery require special skills only known to experts, and new techniques are constantly being evolved and existing ones improved upon. Time-lapse cinematography has allowed highly speeded-up representation of events which normally proceed at too slow a pace to follow their time course, in the striking fashion which this technique permits. Filming of the behaviour of animals in their natural habitat and in their responses to artificially prepared 'dummies' which represent other members of their species or of other species, enables ethologists to study hypothesised genetic and environmental factors in, say, mating behaviour, pursuit of prey, fighting behaviour, and so forth.

Obtaining an electronmicrograph of a synapse between two 'linked' nerve cells (*cf.* Colonnier, 1968 for many beautiful examples, and Pfenninger *et al.*, 1969, for high resolution ultrastructure studies), like the use of other techniques in other kinds of public observations, depends not only on a mastery of these techniques but also on the 'pattern recognising ability' of the human brain, when the latter is used for making interpretative *selections* of observed material. For instance, when examining electronmicrographs of a nerve cell system in order to detect 'synapses' (junctions of neurons), the experimenter must know 'what to look for'. He must select from among his many electronmicrographs of nerve cells those which clearly show what his brain 'recognises' (classifies) as a synapse. Likewise, if an 'elementary particle' physicist wishes to demonstrate the existence of a particular type of track in a photographic emulsion, he may have to scan (automatically, for example) many exposures before he obtains the type of track he is looking for. (For instance, in scattering experiments involving elementary particles, one may have to detect specific, but rare, scattering tracks in order to provide empirical evidence for a particular 'type' of scattering.) Thus 'pattern recognition' or specific *selectivity* of significant patterns (either by use of automated pattern recognisers or by the human brain in its capacity of a pattern recogniser) plays an important role in the observational process.

B. Absence of one–one relations between 'private' experiences and 'public' behaviour signs[1]

Some philosophers of science believe that a distinction between the world of 'private experiences' and that of 'public observations' is artificial, as supposedly some or all private experience can be brought into one–one correspondence with certain human public behaviour (such as spoken or written language, or other forms of symbolic or non-symbolic behaviour).

[1] Lewis (1970, p. 108), in a discussion of some philosophical issues of perception, has disregarded the objections to some of his arguments which are implicit in this section.

For instance, Feigl (1958, p. 398) argued that 'qualitatively identical (indistinguishable) experiences may be had by two or more persons, the experiential events being "private" to each of the distinct persons'. Feigl, however, suggested no standardisation procedures which could assess the identity of private experiences of different individuals (that is, the complete correspondence), and hence nobody could produce any evidence which suggests that this is likely to be the case. That the private experiences of people should exactly correspond amounts to a purely metaphysical statement.

Feigl asked:

> Is it not an 'objective' fact of the world that Eisenhower experienced severe pain when he had his heart attack? Is it not a public item of the world's history that Churchill during a certain speech experienced intense sentiments of indignation and contempt for Hitler? Of course!

Feigl is implying an equivalence (or more precisely a one–one correspondence) of private experiences and of publicly observable behaviour signs which *supposedly* symbolise the private experiences. However, as there exists no independent empirical evidence for any one–one correspondences of private experiences and certain behaviour signs, tacit assumption of such correspondences (or 'equivalences') rests on a purely metaphysical hypothesis. It is not an *objective* (publicly accessible) fact that Eisenhower 'experienced' pain. The old platitude must be iterated: neither his physician nor anyone else in the world felt Eisenhower's pain (see also Hebb, 1954, p. 404). They may have heard his complaints, and other symptoms which are usually taken as indicators of pain. Of course, Feigl fully realises this, but his implied one–one correspondence cannot be substantiated.

In the case of Churchill's sentiment of contempt, people certainly heard his speech, but did not experience *his* private feelings of contempt, although Churchill's speech symbols might have evoked private feelings of contempt for Hitler in his listeners. But this is not the same as saying (a) that Churchill's *feelings* were publicly observable, or (b) that there exists any evidence that the feelings of contempt which he invoked in others were closely similar to his own. Unless one *re*defines publicly observable symptoms of pain (shouts, moans, and so on) or of contempt (such as inflections of voice, and words used) as being logically equivalent to 'private feelings of pain', Feigl's assertions and those of others (including numerous radical behaviourists among psychologists) are not acceptable. Such redefinitions are objectionable on at least two grounds. (1) They imply that behaviour signs are equivalent to the things which they represent (just as if (a) the *word* cabbage were equivalent to the vegetable which it represents, or (b) a library index card were equivalent to the book which it represents). This is a purely formal objection. (2) If we had foolproof evidence that behaviour signs are invariably in strictly one–one correspondence with the private experiences which they are assumed to symbolise, then the classificatory distinction between private experiences and publicly observable (presumptive) symbolisations of these experiences might, indeed, appear as pedantic quibbling.

However, as different people (unless they are monozygotic twins) may contain different alleles (variants) of many corresponding genes, and as our genes control the structure and composition of our brains (*cf.* Wassermann, 1972, 1974), there exists a strong possibility of significant interpersonal variations in those (possibly multimolecular) brain structures whose activated states are (by hypothesis) the physical requirement for the occurrences of specific 'private experiences'. In addition, environmental factors could, during development, cause individual differences in brain structure and/or composition (*cf.* Weiss, 1970). As a result of this, people with different genetic constitutions, and subjected to different environmental factors during development, could have brains which differ in parameters such as biochemical neurospecificity (*cf.* Wassermann, 1972, 1974), and hence produce nerve cells which differ in fine but significant aspects from each other. Brains differing subtly in cellular macromolecular composition could, when peripheral receptor organs are subjected to similar stimulus configurations (and by processed transformations of the brain representations of these configurations), have (epiphenomenally, for example) individually differing 'private experiences' in response to the same environmental stimuli.

In addition people with subtly differing corresponding brain compositions could also have differing degrees of interactions between their various brain structures. Hence, they could not only have different private experiences, but could also produce different publicly observable behaviour symbolisations of their (presumptive) private experiences. As we can neither publicly observe nor compare the 'private experiences' of people of different genetic constitution and environmental histories, but only their behaviour, we could not know whether subjects with possibly subtly different brain constitution, who show similar behaviour responses (for example, when their teeth are drilled without an anaesthetic), have similar or different 'private experiences', as we cannot *directly* compare private experiences (*pace* Burt, 1959, p. 162). Neither could we be sure that different subjects who give different responses to similar stimuli could not have similar 'private experiences'. Thus, one subject might shout three times as loudly as another when subjected to the same pain evoking stimulus, yet, for all we know, the subject who screams loudest might 'privately' experience less pain than the other. We simply cannot know. It is because of these genetic considerations alone that I do not share Feigl's (1958, pp. 398–9 and pp. 429ff) view that 'private experiences' are 'in-principle-intersubjectively confirmable.' (For similar reasons I consider all attempts of philosophers and others—*cf.* Eccles, 1964, p. 272—to discover unique correspondences between 'private experiences' and behaviour as an unrealisable goal.) Thus, Feigl's (1958, p.

429) statement that 'once we have established the laws regarding the correlation of the φ's [= physical body and brain states] with the ψ's [= private experiences] for our own case', then 'indirect verification or confirmation of statements regarding the mental states of other persons . . . is clearly possible', appears to me based on an inappropriate analogy. Accordingly I see no grounds for abolishing the classificatory barrier between private experiences and public observations.

C. Further remarks concerning the metaphysics of realism

In § 1.2D I have already dealt with the philosophical problems concerning the 'reality of the material world'. My views in this and certain other respects are in close agreement with those of Mercier (1970), who believes that questions concerning 'reality status' of things (while of interest to metaphysicians) are officially redundant for scientists. For example, Popper (1970, p. 24) asks:

> Can there be such a thing as a statement or a theory which corresponds to the facts, or which does not correspond to the facts?

In the sense in which Popper uses the word 'fact', a 'fact' must be something which 'exists' absolutely, that is, independently of those statements of theories which are supposed to correspond (or not to correspond) to it. As criteria for ascertaining the absolute 'existence' of 'facts' are not provided by Popper, his answers to his own questions (which depend on the use of metalanguages) are implicit metaphysical hypotheses. Many philosophers and scientists embrace also metaphysical beliefs that there exists a world of 'real systems' (or 'facts') whose properties may be assessed (a) by 'public observations' based on ever improving techniques, carried out by qualified scientists in suitably equipped laboratories, together with (b) ever-changing but successively improving theories. As I pointed out earlier, the metaphysics behind this view is that successive improvements of observational techniques and theories will get us increasingly closer to complete 'knowledge' of the 'facts'. It is a *faith* which many of us share, but which cannot be empirically (and at best historically) defended.

Popper not only applied his metaphysical assertions concerning the existence of 'facts' to hypothetical entities such as elementary particles (*cf.* Popper, 1967; see also p. 62), but he suggested that statements (for example, when written down in books) *per se* are 'facts'. This is apparent from his claim (Popper, 1970, p. 13) that 'The fact that certain theories are incompatible is a logical fact, and holds quite independently of whether or not anybody has noticed or understood this incompatibility.' I pointed out earlier (p. 15) that I do not subscribe to such metaphysical assertions. That theories are incompatible in any absolute sense could not be asserted on the basis of any known decision making procedure. If men ceased to inhabit this earth and scientific theories survived in uninhabited libraries, what significance could one attach to the logical 'compatibilities' of the printed symbol systems which represent the theories? A metaphysical realist could hypothesise that (a) there exist 'real' systems called brains (which he can describe), (b) there exist 'real' systems in the form of printed statements (and the like) which represent theories, and (c) the 'real' systems (a) and (b) can interact, and that this interaction produces statements of evaluation of theories by brains leading to further statements (for example, on paper) that some theories are incompatible. This interpretation is essentially as metaphysical (though more readily understood by non-technical readers) as the statements that (a) there exist 'real' systems called electrons, (b) there exist 'real' systems called protons, and (c) systems (a) and (b) can interact and that this interaction produces certain describable results, which can be recorded by instruments (for example, be printed out by a computer on paper).

The realist argument implies that theories, when existing uninspected by brains on paper, have a 'reality' status in the universe essentially no different from that of elementary particles when not observed by instruments. As some scientists (e.g. Mercier, 1970, and myself) would regard any statements suggesting an absolute 'reality' status of elementary particles (and so on) as metaphysical (although it may *satisfy* metaphysically minded philosophers and scientists), they must on the same grounds regard the allocation of 'reality status' to the 'logical status' of statements stored in libraries (or as inactive memory traces in other people's brains) as being equally metaphysical. A non-metaphysician would argue that the logical incompatibility of statements of certain theories, even if 'inherent' in printed statements (in the same sense that the spin of an electron may be regarded to be one of its inherent properties), can only be ascertained when brains operate on the statements and produce further statements which can again be operated on by other brains. The detection of inconsistencies of, say, written or spoken statements by brains operating on the input which represents the statements may be compared to the detection of properties (such as spin) of elementary particles by suitable observing systems.

2.2 'Public observation' as a classification based on 'property lists'

A. Concepts, properties and valuations

The human capacity to form concepts is a classificatory capacity, and forms presumably an essential function of human brains (on the hypothesis that brains are, among other things, classificatory automata). Man can classify objects and situations and their perceived variants, as well as complex abstract relations. In fact, most, if not all, expressions of a language refer to concepts, that is, to classes of things, although in some cases the class may only contain a single member. Classes can form hierarchies. For instance, when we look at a shopwindow, which contains electric irons, electric blankets, electric convector heaters, arm chairs, tables, dining chairs, and the like, then our brain presumably performs hierarchically ordered classifications. A typical object such as an electric iron consists of several recognisable parts which must bear reasonably 'standardised' relations to each other, if the object is to be 'recognised', that is, classified, as an electric iron (for example, it must have handle, base, flex, plug, and so on, in correct relative positions). Thus, the *subclass* members, that is, the components, must be recognisable if the object 'electric iron' is to be recognised. At least a sizeable number of 'cues' or subclass members must be present to enable recognition of a total object.

Again, electric irons, electric blankets, electric convection heaters (and so on) will be allocated to, say, the second order class (or concept) of 'electric household appliances'. Tables, chairs, and so on, are allocated to the second order class of 'furniture'. Furniture and electric household appliances can be allocated to the third order class (or concept) of 'household goods'. Thus, at each hierarchical level there exists a number of classes and subclasses. Classifications can be concerned with objects and with *apparently* 'continuously' varying situations of a concrete or abstract nature. One can classify a certain physical activity as the concept 'tennis playing', or a particular form of movement of a horse as 'trotting'. Appropriately qualified brains can classify differential equations according to class structure, and the brain of a musician can classify a great variety of compositions, which may differ in detail, as 'fugues'.

The brain (by hypothesis) has the further ability of being able to allocate values to classes, or to specific members of classes, and this property plays a part in the selection of classes or class members of presumed significance, that is, in choice procedures. I am not concerned in this book with possible brain mechanisms which could achieve this, as suggestions concerning these mechanisms form the subject of extensive work now in progress (*cf.* Wassermann, 1974). Although most selections and evaluations and many classifications are at present performed by human brains, it seems likely that some aspects of these activities will in due course be performed by 'artificial intelligence systems' (*cf.* reviews of certain types of artificial intelligence systems by Arbib, 1964; Minsky, 1961, 1967; Nilsson, 1965; Neisser, 1967, chapter 3; Wathen-Dunn, 1967; Zusne, 1970). Several suggested pattern recognisers are based on systems equipped with 'property detectors' (or 'property filters' as they are sometimes called). Property detectors could be of many different kinds. One type could match (say through topological deformations) part or whole of a complete pattern—that is, it could act as a deformable template, which, when fitting part of a pattern, signals this fit to a 'property recorder' system. This, when activated, registers that a certain property has been detected by the system.

Among the most significant known *non*-biological property 'detectors' are man-made instruments. However, human nervous systems, sometimes with, sometimes without the aid of instruments, can classify objects or events. Some neuroscientists have argued that the human nervous system is but a glorified 'property detector' system, the detected properties being referred to as 'features'. However, I have discussed elsewhere (Wassermann, 1974, chapter 3), at some length, that a considerable number of arguments, based partly on experiments and observations, indicate that feature detection is unlikely to provide the means for classification by nervous systems. Human brains can apparently represent and detect complete and complex configurations as *coherent* entities ('images'), and the feature detection method does not explain how images can be coherently represented and how their *complete* configurations and their subconfigurations can be identified and classified (Wassermann, 1974, chapter 3).

The alternative suggestion that brains are endowed with engrams (memory traces), many of which can detect complete configurations and subconfigurations (and not *only* 'features' of configurations), was advocated in considerable detail elsewhere (Wassermann, 1974). In fact, a hierarchy of *engram-mapping* systems could exist, such that the hierarchically lowest order engram-maps match complete configurations or their larger or smaller subconfigurations. At higher order hierarchical levels the successful matching of a configuration of a lower order engram could be indicated in 'token form' (for example, by means of concept-representing entities, such as the concept-representing 'mapping molecules', suggested in Wassermann, 1974, chapter 5). By comparing the same central nervous representation of a particular configuration (whether externally supplied or internally generated as part of 'creative thinking') with many engrams, different ones of these engrams could match different aspects of the same configuration.

Some engrams could match the complete configuration, others could match some specific subconfigurations, while others could match particular 'properties' (such as colours or size present in a visual configuration, or loudness present in an auditory configuration). In this scheme particular concept-representing structures would become activated when one or more specific engrams match a complete configuration, or one of its subconfigurations or some feature of the configuration. Concepts and hierarchies of concepts could themselves be represented on hierarchically higher order engram-maps by concept-representing structures (see above). A configuration formed on one engram-map by a set of concept-representing structures could, in turn, be matched by one or more higher order engram-maps, and be represented by one or more hierarchically higher-order concept-representing structures on a higher order concept-representing engram-map, and so forth.

The procedure just described, while starting often, but not necessarily always, with an explicit representation of each configuration, could permit conceptual analysis of complex configurations. Thus, this hypothetical procedure allows brains to provide concrete as well as conceptual representations. In contrast to this, artificial pattern recognisers do not have to represent configurations explicitly (that is, in 'concrete' form) within the automaton concerned, but could immediately proceed to detect properties, that is, they do not require to form *coherent* representations first. Properties can range from simple to highly complex ones. Typical 'simple' properties of geometrical configurations could be (a) colours, (b) shapes, (c) topological features, such as connectivity properties, (d) properties of spatial order such as 'included in', 'larger than', 'to the right of', 'above', or 'behind',[2] (e) metrical properties, for instance rectilinearity, triangularity or circularity, (f) other geometrical properties such as convexities of contours of figures, (g) distances intercepted between contours. One could also detect properties relating to events extending over time intervals. Examples of such properties are (h) relations, such as 'earlier than', or (i) detection of movements of components of systems relative to each other. At a higher conceptual level one could detect properties of *abstract* relations (such as the characteristic structures of certain types of ordinary algebraic equations, which allow one to classify their structure as, say, quadratic or cubic equations). Such property lists could be extended indefinitely.

I must emphasise that although a property-detecting system could detect the coherence of a configuration as a *property* of the system, this is not the same as representing the configuration itself in explicit coherent form within the system, as appears to be done by brains.

Certain pattern recognisers, based on 'property filters' (see p. 47) use 'property lists' (*cf.* reviews by Minsky, 1959, 1961; Hawkins, 1961). According to Minsky (1961, p. 12) a pattern recogniser based on property list procedures can in *simple cases* recognise, say, n (where n is fixed) different properties $P_1, P_2 \ldots P_n$ of patterns, which can be enumerated by single integer suffixes. (In contrast to this the human brain can form an indefinite number of engrams, including engram representations of concepts.)

A property list-using automaton can characterise any pattern according to the presence or absence of each typical property P_i which the system is able to detect on the assumption that the system always functions to maximum capacity. (In contrast to this, human brains often function suboptimally, by failing to recognise things which they were previously able to recognise, such as a piece of music.) As mentioned, the suffix i of property P_i may in simple cases be an integer. Although a property, if present, is signalled by the property filter *output* as a *single* invariant, the property filter could *respond* to a limited *range of variants* of the property, such that each variant of the

[2] Some of the listed properties can be interpreted in the language of partially ordered mathematical sets (for axioms, *cf.* Bunge, 1967c, vol. 1, p. 414).

same property, elicits the same output signal of the filter. In more general cases the property may vary with time (t), and also depend on several variables of the sampled system, and the suffix i of P_i could then be characterised in the form

$$i = [j(t), j(t_0)] \tag{2.2.1}$$

where $j(t)$ denotes the values of the set of state variables

$$j(t) = [i_1(t), i_2(t) \cdots i_r(t)] \tag{2.2.2}$$

In (2.2.2) the value of the suffix r depends on the property concerned, and in (2.2.1) t_0 refers to an initial time.

Although in some scientific *theories* certain abstracted 'properties' of systems are *hypothesised* to be continuous, and are represented by continuous mathematical variables (for example, by use of curve fitting procedures which involve numerous hypotheses; see § 2.4), this does not imply that the 'corresponding' empirical 'properties' possess 'empirical continuity'. In my opinion 'empirical continuity' is a metaphysical concept, for several reasons. (1) Inexactness of all measurements and observations would exclude verification of 'empirical continuity' even if it should exist and if it corresponded to mathematical continuity. (But, for all we know, even physical space and time might ultimately be quantized; *cf. Nature*, 1973, **246,** 378.) (2) Only *mathematical* continuity is *definable*, while 'empirical continuity' has (within science) never been given an empirically anchored definition (as distinct from *theoretical*—that is, hypothetical—continuity). To the *naive* molar observer ('the man in the street') water may appear continuous, and, in accordance with this crude observation, continuum mechanists have established *theories* which postulate a continuous density. However, experimental techniques based on X-ray diffraction methods (*cf.* H. S. Frank, 1970, and F. Frank, 1973) suggest that water has a discontinuous structure approaching that of ice I—that is, X-ray data provide no evidence for 'empirical continuity' of water in any *definable* sense.

Again, if somebody talks about a 'continuous spectrum' of radiation being emitted by the atoms of a gas, this 'continuity' is only used for classificatory purposes at a certain level of observation. In fact, if a photographic negative of a 'continuous spectrum' is being sufficiently enlarged, the grains of the photographic emulsion may become visible, and they do not form a 'continuum'. It is because of examples like these (and innumerable others) that I do not agree with Butchvarov (1968), who claims that 'continuity is an empirical fact', and who endorses Körner's (1966, p. 49) view that (to cite Butchvarov) 'the mathematical theories of continuity are at best analyses of mathematical substitutes for this empirical fact'. As far as I can see neither Butchvarov nor Körner have demonstrated that continuity is an *empirical fact*. (*Hypothesised* continuity certainly exists in many theories.)

In some cases a single state variable may be assigned a constant value for an interval over which a 'property' varies. For example, when a monatomic gas is being excited and its atoms make radiative state transitions, then the frequencies of certain spectral lines may be regarded as 'properties' ν_i, where i is of the form (2.2.1) and t in (2.2.1) is large enough to ensure that radiative transitions have occurred (*cf.* Condon and Shortley, 1935, for a detailed discussion of the spectroscopy of complex systems). In these circumstances the set i may be interpreted as referring to a narrow empirically observed interval over which the values of ν_i, may range on account of 'broadening' of spectral lines (through collisions between gas atoms).

In order to characterise the presence of a property in a sampled pattern a 'measure' function f_i is being associated with property P_i such that f_i can assume two values '1' and '0', and

$$f_i = \begin{cases} 1 \text{ if property } P_i \text{ is present in the sampled} \\ \text{pattern or observation} \\ 0 \text{ if this is not the case} \end{cases} \tag{2.2.3}$$

A pattern or observation can be characterised by a vector

$$\boldsymbol{f} = (f_1, f_2, \ldots f_n) \tag{2.2.4}$$

called the *character* of the pattern or observation. If there is only a single and enumerable state variable, then the suffixes in (2.2.4) can be regarded as representing the state (or 'property') enumerators and the character vector is n-dimensional. In the more general case represented by (2.2.1) and (2.2.2) the dimensions of the (hypothesised) space might be non-enumerable. The vector space spanned by the characters of the system is called the *property space*. (In the simplest case when i in (2.2.1) is independent of time and can be represented by an integer, the property space is spanned by 2^n characters.) General features of simple property spaces were discussed by Farley (1960). He also dealt with *hierarchies* of property list procedures, which could account for a hierarchy of classifications of a pattern by a pattern recognising system.

In the immediately preceding passages I have dealt with property detection and property lists, which are based on 'artificial intelligence' (= 'machine intelligence') systems. In such systems the 'property detectors' immediately abstract properties from the *externally* given configuration. In other words, these artificial systems do not first provide an internal representation of the configuration, and use this as the starting point of property detection. I argued earlier in this section that, in contrast to this, human brains seem to provide complete internal 'images' of external configurations, and then use these 'images' as starting points of classifications, which are based on the establishment of concept-representing structures. Accordingly we may expect that in 'public observation' the human nervous system does not function like a property detecting automaton, but relies on the brain's dual capacity for (1) internal representation of perceived patterns and (2) classification of these patterns in terms of concepts.

For instance, detectable cells of an organism can, with the aid of instruments, be classified by the human brain according to shape, size, chemical composition (and so on) into cell classes (such as osteoblasts of bones, nerve cells, muscle cells). In turn, subcomponents of these cells can be classified according to detectable patterns, into postulated structures, such as the 'endoplasmic reticulum', the 'cell nucleus', 'mitochondria', 'ribosomes' and so forth. The typical class patterns are illustrated and interpreted with the help of many electronmicrographs in numerous papers and textbooks (*cf.* Miller, 1973, for an electronmicrographic illustration of bacterial multiple ribosomes, so-called 'polysomes').

In other cases the properties of systems which human brains can classify may depend on measurements. Thus, for a gas, its volume, temperature and pressure form measurable properties, and the brain patterns and their classification which arises in such cases are related to inspection of measuring instruments. ('Measuring techniques' would nowadays include such things as chromatography, and other procedures, which show a variety of types of precision.)

The various concepts which assist in the classification of measuring procedures permit us to schematise 'public observations' at coarse as well as at sophisticated levels of observation. However, *it is important to distinguish observed from postulated properties* (see §§ 2.3 and 2.4). (For a different but interesting approach, the reader is referred to Körner (1966), chapters 1 and 2.)

B. 'Physical things'

While many philosophers and scientists agree that science is concerned with publicly observable properties of systems, it has been argued that without man, who is a *conscious being*, public observation would not be possible. The emphasis of the argument is that man's consciousness is somehow a requirement for his scientific, as well as for many of his other intellectual activities. However to argue along such lines, or to suggest (at the other extreme) that all public observations could *in principle* be accomplished by physical recording systems (Braithwaite, 1953, p. 7), is purely metaphysical, since it can neither be refuted nor confirmed by any conceivable means. Even if all laboratory work could be taken over by artificial intelligences, it could still be argued that conscious man originally designed these, and that his consciousness was instrumental in producing instruments and artificial intelligences. Those who make such claims cannot substantiate them, nor as far as I can see, could any counterclaims be convincingly established.

Advocates of the counterclaim would have to provide evidence (a) that there exist other structures in this or other galaxies in the universe, which are 'inhabited' by unconscious robots, (b) that these robots are *genuinely* unconscious (which would always be hypothetical as they might be conscious but try to pretend that they are not), (c) that these supposed unconscious robots had evolved without the help of conscious beings, and (d) that they had built scientific laboratories and made significant discoveries.

For the scientist it is a sufficient hypothesis that human brains can act as pattern recognising automata and can with the aid of man-made instruments (which are consequences of the inventive capacity of brains) detect physical properties. For this purpose one has ultimately to hypothesise how brains (operating as automata) could act as pattern recognising systems (and as creative, that is, inventive, systems), but this is the concern of further work now in progress, and not the subject of the present monograph. If one can suggest a set of *hypotheses*, consistent with empirical evidence, indicating how brains could recognise (and recall) patterns, invent scientific machinery, plan experiments (and so on), then this will amount to another aspect of scientific theorising (see below) and is in line with the view that all scientific hypotheses always remain unprovable assumptions, which may either be further confirmed or ultimately refuted.

Hypotheses as to how human brains could generate hypotheses (as well as all forms of other creative activity) would serve to dispense with the need to regard man as a black box. This would have the advantage of producing a *closed* hypothetical system, that is, a scientific theory of generalised knowledge acquisition, all *generalised* knowledge consisting of hypotheses. It must be stressed that *generalised* knowledge refers to scientific knowledge (including creation of hypotheses) and not to idiosyncratic knowledge (such as knowledge of a particular symphony or of a particular poem), so that the word 'knowledge' functions in the present context in a quite specific and restricted sense. It does not designate mere acquaintance, but formulation and evaluation of hypotheses. (However, an adequate theory of brain function must also be able to explain how brains *via* their structures and processes can store and generate idiosyncratic knowledge.) There exists no evidence that it is the consciousness associated with brain activity which is responsible for achieving scientific observations (or for inventing observing machinery). But it is up to brain-model theorists to explain how postulated brain machinery could perform all the required transactions on its own, so that interactionism becomes a redundant (and uneconomical) hypothesis.

For present purposes it is sufficient to leave aside all discussions as to how *brains*, functioning as mere automata, could design laboratory machinery, design (that is, plan) experiments, detect 'properties' of systems and use these to build up theories. One can define a 'physical thing' as one which has at least *some* publicly observable properties, detectable by brains aided by observing systems. Taken from this point of view, the human brain is itself a 'physical thing', while its 'private experiences' are not 'physical things'. The publicly observed properties of a 'physical thing' which are known at any one time may not exhaust the whole range of its observable properties, and we can never know when we have exhausted these. Moreover, properties are not the same as entities.

For instance, many properties of a hypothesised system cannot be observed simultaneously, as testing for different properties may require different types of observing instruments, which might not be applied to the observed system simultaneously for practical reasons. When physicists talk about elementary particles (such as electrons, positrons, or neutrinos) they cannot claim to have observed such particles as 'entities', but only to have discovered evidence consistent with and confirming certain *postulated properties*, and the set of these properties collectively defines what they mean by a particular 'particle'. Similarly, when quantum chemists talk about atoms, they do not claim to have observed these directly as entities, but to have obtained evidence for numerous postulated properties of the atoms and their subconstituents (electrons and nuclei). They may postulate for 'electrons' the properties of electron spins, the property that 'electrons' satisfy Schrödinger's equations, that they satisfy the Pauli exclusion principle, and so forth. (These properties are, of course, in the case just discussed postulated in terms of hypotheses of a theory and not derived from direct public observations.)

Thus, scientists are ultimately only concerned either with observable or postulated *properties* (including patterns) of systems, but not with 'systems' or entities *per se*. The 'structure' (and postulated function) of haemoglobin (*cf.* Perutz, 1969, 1970) is not based on the direct observation of molecules but on the interpretation (involving massive computations) of X-ray diffraction data, combined with the theory of X-ray diffraction. Anatomical descriptions of animals or plants amount to descriptions of class properties which may include structural properties (patterns and form). If I inspect a photo of a giraffe I can identify the latter rapidly, because (by hypothesis) my brain classifies its relevant external configurational details, thereby activating the concept-representing engram structure 'giraffe'.

Several philosophers of science have not fully appreciated that many human statements are used in science in order to *classify* a property. Even in cases where such statements *refer* to a private experience, the scientist is not concerned with the private experience *per se* (*pace* Woodger, 1952, p. 283). For instance, when Penfield and Rasmussen (1950) reported that a human subject, when electrically stimulated at a specified cerebral locus, announced 'I see red', the word 'red' signifies the property that this type of stimulation leads to a specific class of verbal responses, namely responses that could also be obtained from a non-colourblind subject when retinally stimulated with light falling within a certain range of wavelengths. In fact, if Penfield and Rasmussen had both been colourblind and the word 'red' had no 'private experience' significance for them, the equal classification of the two types of responses (to cortical stimulation and retinal stimulation with red light respectively) would still have had the same implications (see also Feigl, 1958, p. 385 for a similar point of view).

C. Phenomenalism and Geisteswissenschaft

Some 'phenomenalist' philosophers put forward the metaphysical assertion that *verification* of the 'existence' of physical objects depends on the use of 'sense data' (that is, on a set of 'private experiences'). Adherents of this view argued that all public observations depend directly or indirectly on 'private experiences', and this type of phenomenalism tacitly implied at least a one-sided psychophysical 'interactionism'. It is this hypothesis which underlies the metaphysical assertion that consciousness plays a relevant part in scientific observations. I stressed already that improved neuropsychological theories are likely to show that this metaphysical argument is redundant. The metaphysical argument was clearly stated by Bertrand Russell in 1917 (see Russell, 1953, p. 140):

> In physics as commonly set forth, sense-data appear as functions of physical objects: when such-and-such waves impinge upon the eye, we see such-and-such colours, and so on. But the waves are in fact inferred from the colours, not vice versa. Physics cannot be regarded as validly based upon empirical data until the waves have been expressed as functions of the colours and other sense-data.

The fact that non-colourblind observers will normally *name* light of a particular wavelength correctly does not imply that conscious states were causes of their verbal responses.[3] The wavelengths are observationally related to a public statement but not to a private experience. On the other hand interactionists could argue that interpretation of spectrographs requires consciousness of brains. But this would only shift the metaphysical argument further back. The only way to demonstrate the redundancy of interactionism, is to show in terms of a neuropsychological theory that brains working as pure automata (admittedly conscious automata) could accomplish all scientific transactions, not as a result of the occurrence of conscious experiences, but through machine-like operations only. No amount of inspection of works produced by man could help to confirm or refute interactionist claims. If someone doubts whether Shakespeare wrote his works as a consequence of being a conscious human being, or whether he produced these by purely automatic activities of his brain (although these

[3] Russell's arguments are partly inappropriate. For most purposes spectroscopy could be carried out by colourblind people. For instance, if a gas of known chemical composition emits light its spectrograph can be determined by measurement, irrespective of verbal statements of the colours concerned. In fact, spectroscopic analysis of the light emitted by distant stars or by gases of unknown composition in the laboratory allows one often to identify the composition of gases in a manner which is not related to the human capacity for colour vision.

automatic activities could well have been *accompanied*, epiphenomenally, by conscious experiences), this cannot be decided by inspection of his plays. Only appropriate brain-models, based on known or hypothesised functions of nervous systems could suggest to what extent purely automatic brain functions could accomplish some of the most highly sophisticated tasks. Accordingly, claims like those by Walshe (1953), that only conscious beings could produce sonnets, are tantamount to the premature suggestion that we have already attained *complete* knowledge of what brains, working as automata only, could accomplish. The possibility that an automaton could perform a highly sophisticated activity (such as writing a sonnet) *and* be conscious, does not imply that the automaton's consciousness must be a cause of this activity. (A relevant discussion, differing from my own in many points, was given by Ayer, 1956, chapter 5.)

I shall exclude from science the so-called *Geisteswissenschaften*. Eysenck (1961) referred to the latter in the following passage:

> What is said is something like this. Psychology by its very nature cannot be a *Naturwissenschaft*, i.e. a natural science like physics or physiology, but must be a *Geisteswissenschaft*, i.e. a kind of intuitive, humanistic discipline; that psychology cannot *explain* behaviour in terms of general laws, but can only *understand* it in terms of each individual's own intuitions. This is a line taken, among others, by Husserl and other German philosophers whose *a priori ex-cathedra obiter dicta* have attracted far more attention than their lack of factual information of modern psychology and its ways of working would seem to warrant.

2.3 Operational specifications of public observations

A. 'Operationalism'

By definition, all publicly observable systems must have *some* properties which can be directly specified in terms of recordable operations of instruments and/or describable human inspection (Bridgman, 1927). To this are usually added *interpretative arguments* which give meaning to the recorded observations (for example, an interpretation may state that a certain structure in an electronmicrograph represents a mitochondrion). Many scientists and philosophers of science who say that theories must be 'empirically anchored' wish to imply simply that some hypothetical properties of certain systems or events are indirectly related to public observations. In § 2.4 I shall discuss some of the simplest of these relations. However, this is not remotely the same as the extremist notion that *all* properties of a system (in particular intricate hypothesised properties) must be *directly* expressible in terms of laboratory observations. Alas, some 'operationalists', notably some radical behaviourist psychologists, subscribed to these extreme views. For instance, at one time Skinner (1947, p. 28; also cited by Verplanck, 1954) claimed that theories

> are statements about organizations of facts . . . . They are all statements about facts, and with proper operational care they need be nothing more than that.

(For Skinner a 'fact' is a directly observable property of a system.) Skinner's belief is rooted in the solemn conviction, popular at the end of the last century, that theories should only refer to, and relate to, properties which can be publicly observed directly. It is a belief which disregards the evidence that even the simplest laws or empirical generalisations are *hypotheses* (§ 2.4) and *not* facts.

The nineteenth-century 'natural philosophy' which opted for theories based exclusively on relations between 'immediately observable properties' derived from Mach's extremist philosophy. In physics, thermodynamics formed its paradigm (see Bunge, 1967a, pp. 1–6, for important relevant remarks). Equations of state of gases (such as van der Waals' equation), Ohm's law, and numerous other simple 'molar laws' (see below) are only based on theoretical variables which are immediately extrapolated from publicly observable experimental variables. This may have misled some scientists who received only a basic school training in physics into believing that these types of 'laws' are characteristic of physical theories. But this is plainly not the case (see below).

Some 'operationalists' were probably influenced by metaphysical considerations, concerning the 'reality status' of certain postulated microsystems. Without valid justification they argued that any theoretical variable of a theory should be *directly* related to sets of observable data. This would be somewhat as if tailors had decided that women must only wear men's clothing. Yet it soon turned out that in many sophisticated theories (such as quantum mechanics, and optics based on Maxwell's electrodynamic theory) this goal is utopian, and many theoretical variables (the so-called 'unobservables') and many hypothesised theoretical properties of systems are only very indirectly related to publicly observable properties. For instance, in order to apply Maxwell's theory to optical imaging, complicated statistical averagings over space and time parameters of 'unobservable' theoretical variables which are not simply related to observations (that is, the postulated electric and magnetic field variables) may be required in order to obtain results which can be compared with observations (*cf.* Theimer *et al.*, 1952). These time and space averages may not contain the 'unobservables' in a form which allows 'inversions' of the results. Hence, while those theoretical variables which generalise from experimental variables can in this case be expressed as averages over unobservable variables, the resulting relations of the theory cannot be solved so as to express the unobservable variables in terms of the observables. (This also creates difficulties for Popper's falsifiability view of theories.)

A closely related problem appears also in various attempts to deduce non-equilibrium (as well as equilibrium) thermodynamics from quantum mechanics. Although thermodynamical variables can be expressed in terms of quantum mechanical variables, the statistical averaging processes do not allow an inversion of the calculations; that is, quantum mechanical variables cannot be expressed in terms of thermodynamical variables only. Hence, even if people should succeed in 'deducing' the main results of non-equilibrium thermodynamics on a rigorous basis from quantum mechanics (*cf.* Born and Green, 1949, for some important suggested formulations), this would not help to express quantum mechanical variables (such as Schrödinger functions) in terms of those variables of thermodynamics which can be immediately related to observations.

A powerful example which illustrates the manner in which *hypothesised* properties and postulated 'unobservable variables' (such as Schrödinger functions) enter into a complex theory, is provided by the deduction of Ohm's law in Bloch's quantum theory of electric conduction in metals (*cf.* Wilson, 1958; Peierls, 1955; Harrison, 1970). To start with, the theory hypothesises a complex metallic ionic lattice, consisting of metallic (for example, copper) ions located at hypothesised loci, which can be independently related to a theory of X-ray diffraction studies of metallic structures. The ions are assumed to make small 'thermal' oscillations about their mean positions, being in first approximation harmonically coupled. (Anharmonic coupling enables thermal energy transport, thereby allowing excess accumulation of energy gain by the ions to be conducted towards the boundary of the metal.) Next, the Bloch theory of metallic conduction postulates an 'electron gas' which, under the influence of the electric field, drifts through the ionic lattice. The approximate quantum mechanical treatment of the process is based on (a) a postulated 'Hamiltonian function' (Hamiltonian) of the lattice, and (b) a postulated Hamiltonian of the electrons which includes appropriate coupling terms with the lattice ions.

This part of the theory, making allowances for the Pauli exclusion principle, permits (on the basis of numerous hypotheses concerning approximations in the interaction formalism) calculation of the probabilities that individual electrons gain a certain amount of momentum from, or lose a certain amount of momentum to, the lattice ions. Making use of these probabilities one can establish Boltzmann's equation for the total 'electron gas' which, under the influence of the electric field moves through the ionic lattice. The approximate solution of this equation, jointly with further statistical calculations, allows one to deduce Ohm's law in good approximation. Inspection of the calculations (*cf.* Wilson, 1958; Peierls, 1955; Harrison, 1970) shows that the numerous computational steps relate the quantum mechanical variables of the earliest stages of the theory only very indirectly through non-invertible averagings with the molar variables (current, voltage) and with the 'resistance' which appear in Ohm's law. Many similar examples in which quantum mechanics, *via* statistical mechanics, is being related to observations (*cf.* Fowler and Guggenheim, 1939; Fowler, 1936) could reinforce the preceding arguments regarding the non-invertibility of steps leading from quantum mechanical properties to 'molar' observables. (See my remarks on p. 57 concerning Popper.)

B. Operational specifications

An experimentalist has to describe carefully the type of system he is concerned with and the methods used in preparing the system and in observing its properties. Inspection of typical papers on molecular biology in the *Proceedings of the U.S. National Academy of Science* or the *Journal of Molecular Biology*, will provide the reader with good examples of statements of techniques and material used, and so on. Preferably all conditions which are *believed* to affect the outcome of an experiment should be stated. These conditions are often not known until many experiments have already been performed, which gradually suggest that 'factors' affecting the experimental outcome are present, which were previously disregarded or unknown. Hence, a successive 'tightening up' of experimental conditions may take place in the light of new experience gained. For example, in the early experiments concerned with enzyme induction, some experimenters were not sufficiently careful to guard against the possibility of mutant strain selection, and this possibility was only eliminated in later, more carefully controlled, experiments.

A danger inherent in many experiments, notably in molecular biology, is the possibility of experimental 'artefact'. One is dealing with immensely complicated systems, which can be affected by numerous factors not readily controllable by the experimenter. For instance, on isolating cell membranes one may find certain proteins in the membrane. Does this mean that these proteins are normally present in the living membrane, or that they were adsorbed by the membranes during isolation and preparation of the latter for examination? The only answer to the critical battle-cry 'artefact' is the design of new and better experiments which make interpretations in terms of certain artefacts implausible, even if the artefact cannot be completely eliminated.

Observing instruments and specific material used are nowadays often supplied by manufacturers, and reference to their literature on the specification of the instruments or materials, or *personal* acquaintance with similar preparations and instruments made by other manufacturers, provides sufficient information for the expert. If discrepancies of results are reported from different laboratories, one must make sure that the samples of material used (their purity, for example) were closely similar, that the instruments used were designed so as to provide similar observations of the material used, and above all that the experimental techniques used in preparing the materials were similar. For instance, in preparing cell membranes for examination of their composition, several different techniques have been used (*cf.* Wassermann, 1972, chapter 3) and disagreements in interpretations are sometimes believed to result from subjecting membranes to different methods of preparation for observation.

Instrumental operations, and procedures used in experiments, are not described in detail in publications, if they are of a standard type familiar to experts. People often refer to previous publications from their own laboratories, or those of others, which contain detailed accounts of various techniques.

Many observations involve comparisons of observed properties with certain 'standards'. The 'standard' could be a *measured* quantity (length, mass, time interval) given by an instrument whose measuring scale has been graduated by comparison with an international standard. In other cases the 'standard' could be simply a brain engram. If, for example, a student of cellular ultrastructure states that a certain structure in a certain electronmicrograph is a synapse, he may do so because his brain has formed an engram of previously inspected electronmicrographs of synapses (just as a brain recognises a particular friend because it has formed an engram of his facial appearance). Many scientists could without resort to other electronmicrographs immediately 'recognise' specific ultrastructural features, because their brain contains the required engram 'standards' (just as an experienced X-ray diagnostician in a hospital can detect a bone fracture without resort to comparison with other X-ray plates). In fact a novel feature is often detected if a scientist on inspecting, say, an electronmicrograph discovers something he (or anybody else) has never observed before.

Let us consider another example. Leder and Nirenberg (1964a, b; see also description in Wassermann, 1972, p. 148) labelled *specific* aminoacyl transfer RNAs (= a-tRNAs) with radioactive components, and observed those synthetic triribonucleotides, *of known composition*, which accepted the labelled a-tRNAs, and with the help of this removed some remaining uncertainties of (and also strongly confirmed) the genetic code. The standard system used here for comparison were the synthetic triribonucleotides, whose composition was assured by their method of preparation. In another example, already cited (p. 54), a verbal response to cortical electrostimulation was *compared* with a response (usually) made on request when a normal subject is stimulated with light of specified spectral composition.

C. Empirical statements

Empirical statements describe properties detected by public observations, and are not concerned with hypothetical properties attributed to systems within theories. Schlick (1925, p. 68) argued that all analytical propositions of science are *definitions* and all synthetic propositions are *hypotheses*, because

> all facts of the past even of the instant just passed, are without exception in principle only inferable; the claim that they *have* once been observed might be the result of a dream or of a

deception. For this reason, taken from the point of pedantic accuracy, even historical propositions have the character of hypotheses.

The weakness of this argument lies in a misconception of the assessment of evidence. If I am in possession of an electronmicrograph of a mitochondrion or other cell constituent, or an X-ray of a patient's ulna, then these remain the same factual records—that is, the same observational material—that they were when first recorded (unless they have deteriorated, or unless there exist reasons to believe that they were faked). Schlick would be right in saying that we can only infer any claim that these records were taken *at a certain date*. But the records, and the date when they were taken, are not the same. In the case of historical statements this might indeed have serious repercussions, as the date when a particular recorded entry was made could be very significant. It is also true that in certain scientific observations dates of recordings might be very relevant (for example, when studying the temporal progress of some state of affairs, such as embryological development, the time changes of a geological 'fault' in the earth crust, the time course of biological rhythms, or the intervals between recordable high energy emission by pulsars). Some of these recordings can be done with automatic recording devices and the periodicities of the events can be studied long after these events have occurred.

A scientist's initial task is to *classify* publicly observable properties of systems. The empirical statements of science, as distinct from those of history, are not concerned with unique events but with the properties which can be, or have been, repeatedly observed in closely similar (or at least related) form, that is, properties which are variant members of a *property class*. There are variant classes for many different properties. If *E. coli* bacteria of a specific strain are being repeatedly subjected to observations, the properties assignable to the same specific classes may be observed for many members of that strain. It is a significant feature of the universe, and one which alone makes science possible, that many things occur in publicly observable, closely similar multiple copies and that many phenomena are repeatably observable in closely similar form.

If phenomena and things each occurred uniquely to an extent where no thing in the universe resembled any other, and no phenomenon were even approximately repeated, the world and its events would be chaotic mosaics. Instead, similar class properties of, for instance, leaves and stems of innumerable oak trees permit botanical classifications of oak trees, and similar remarks apply to vast numbers of other organisms. The properties of inanimate systems are also similar for many different samples (subject to variations in impurities, and so on). The periodically repeated planetary movements round the sun, and, more generally, the repeat appearance of various astronomical systems of different orders of magnitude (such as quasars, galaxies, pulsars or supernovae), or, at the other extreme, the normal appearance of DNA in the chromosomes of all (eukaryotic) animal and plant species studied, bear witness that science is possible because properties appear repeatedly. This applies to molar as well as to hypothesised molecular atomic and subatomic systems (although in the latter case presumed repetitions of properties, for example of many similar molecules of a gas, are 'strictly' hypothetical),[4] that is, repeatable properties occur at different postulated hierarchical levels.

The latter circumstance has made it possible to introduce into science hierarchies of systems, many being partly of extremely hypothetical types, defined by hypothetical properties which are only very indirectly related to public observations (*cf.* pp. 57–59). This has greatly increased the complexity as well as the progress of science. Even if one does not join the realists among quantum mechanists (*cf.* the volume edited by Bunge, 1967a), by assigning reality status to 'entities' or 'microsystems' (such as atoms or elementary particles) one can nevertheless reject the positivist view that science should only deal with direct relations between publicly observable properties. Reality for metaphysicians refers to the 'existence' of something. A realist could hypothesise *properties* of 'microsystems' and treat these systems (defined by means of their set of hypothetical properties) as if they existed.

(I suspect that Popper and others who are quantum mechanical 'realists' have something like this in mind. They resent positivist dogma which asserts that only relations between publicly observable properties should have a place in science. In practice—see above—this does not happen, in any case, in many branches of physics.) One can, however, accept all existing statements about elementary particles and their hypothesised fields, and also (tentatively) various aspects of quantum field theories, without calling oneself a 'realist' as far as these systems go. We neither know whether the existing hypotheses are complete (a point often made by various philosophers of science) nor whether they do not in parts require radical modifications or replacement (see E. L. in *Nature*, 1972, **236,** pp. 194–5).

The crux of the matter is that certain empirical statements may be *interpreted* as referring to hypothetical properties of postulated 'microsystems'. For instance, spectral lines of hydrogen may be attributed to electrons of hydrogen atoms, and the fine structure of such spectral lines may be attributed to properties of postulated 'electron spin' (Dirac, 1958). Whether the electrons 'exist', or whether DNA of *E. coli* 'exists' in the sense in which a tree may be said to 'exist' is really a totally irrelevant issue. What matters is that the hierarchy of postulated systems and microsystems (each postulated in terms of its properties) allows us to create a great explanatory coherence among phenomena (see below). If we say that electrons exist in

[4] It should be remembered that although 'molar' systems can often be 'directly' observed (the sun or the moon, for example), they are in 'reality status' as hypothetical as atoms, electrons (and the like) which cannot be 'directly' seen.

many copies in the universe (and likewise many other elementary particles) we do not mean to imply that we can observe these as real entities. The statement only means that the consequences of postulated properties subsumed under the generic term 'electron' can be observed in many places on earth, when many different molar systems are being publicly observed by specified procedures.

Empirical statements refer to observed properties under specified experimental (or alternatively uncontrolled) conditions. In the case of *uncontrolled* public observations, the primary phenomena studied (such as thunder and lightning, earthquakes, volcanic eruptions, the behaviour of animals in their habitats, cases of sickle-celled anaemia and other clinical observations) are not induced by an experimenter. In *controlled* observations the experimenter decides and fixes some properties of the observed system—that is, he 'controls' the state of the system to some extent. For instance, when working with (hypothesised) atomic systems (which are described by quantum mechanics) the experimenter manipulates the experimental machinery and thereby (according to quantum mechanical *theory*) imposes 'perturbations' on the observed system (for example, on a set of similar hypothesised atoms in a container). According to quantum mechanical theory this manipulation puts the atoms into a superposition of particular physical states at a particular time. The *interpretation* of the outcome of experiments in such terms depends therefore on a complicated theory of how 'perturbations' (caused by instruments) could affect hypothesised atomic systems.

As spectral lines emitted by specific substances (and likewise the famous Ritz combination principle) were known long before quantum mechanics, public observations of spectral lines *per se* do not depend on their interpretations. However, often experiments are performed in order to test or confirm a particular interpretation. In such cases the total experimental system and its manipulation are *theory-orientated*. This, for example, is the case in experiments specially designed to test deductions from quantum mechanics.

In many sciences observed properties of typical individual members of classes are in elementary textbooks described by diagrams and statements which represent extrapolations. For example, descriptions and diagrams found in some textbooks of human anatomy represent extrapolated statements which do not refer to a specific (and unique) male or female of the human species. In any one individual slight or considerable deviations from this generalised description may be present. Generalised accounts are equivalent to hypotheses. In fact, generalised statements of less sophisticated biological textbooks harbour the danger of concealing the possibly significant degree of variation in structural and functional detail (cell numbers, and so on) between different members of the same species.

Some publicly observable properties of certain systems can be specified numerically in terms of empirical variables. It is for some purposes profitable to regard numbers as symbols of a specific type of language (namely arithmetic), thereby removing the somewhat artificial distinction between 'descriptive' accounts (using ordinary language symbols) and numerical specifications. Even biology, which until recently was largely descriptive, is showing more and more quantitative features (for example, at the level of molecular biology and biochemistry, as well as at the molar level, where cell counts of particular tissues are assuming importance). Some observable properties are of a *relational* form. This applies to topological properties of many biological systems (for example, statements describing which parts of a brain are *connected* by particular nerve tracks). Likewise experiments of the type carried out by Penfield and Rasmussen (see p. 54) in which the human cerebral cortex of patients was stimulated, provide examples of *relations* between a subject's verbal responses and the loci of cortical stimulation (for example, when the subject described a particular hallucinatory experience, while being stimulated at a particular cortical locus). It must also be borne in mind that many properties which at one time could not be numerically described have since become amenable to numerical specifications (*cf.* Zusne's (1970) discussion of stimulus variable in the human visual perception of form).

D. *'Variables'*

As noted above, publicly observable properties can often be characterised by numerical values of one or more mathematical 'variables'. The degree of precision with which values of experimental variables are stated depends on the type of experiment concerned. Thus, if desired, ripe seed colour of a pea variety could be referred to a numerical colour scale, or be stated in terms of spectral colour composition. Yet Mendel did not resort to such delicate measurements, but preferred verbal statements of the colour, which for practical purposes were sufficient. Again, the words of a language could be regarded as values of a 'variable' in a verbal test (for example, if a subject is asked to name all objects which he can think of that belong to the class of furniture).

The operational specification of an experimental variable sometimes *appears* to depend on a *theory* of function of the instrument used in recording the values of that variable (*cf.* Duhem, 1954, p. 154).[5] For example, the use of a tangent galvanometer for measuring electric currents involves a *formula* based on electromagnetic theory. However, an experimenter who is ignorant

[5] In many cases the theory of an instrument is very complex and unessential for the purpose of observations. While the instrument user ultimately wishes to understand how an instrument records or produces a particular effect, instruments are often successfully used before their operative theories are fully understood. This applies, for instance, to lasers, whose theory is highly complex, but not essential for the user of lasers. (See Haken, 1970, for an account of the theory of lasers.)

of the method of its derivation can use the formula as part of the total operational procedure. He can simply use the formula as an operational tool in a 'paper-and-pencil operation' (*cf.* Bridgman, 1927), for *defining* currents in terms of needle deflections of the galvanometer. Similar remarks could apply to many other experiments where the use of formulae could be regarded as providing simply a *definition* of a variable in terms of observed instrumental readings.

Rozeboom (1956) provided an extensive discussion of the concept 'variable', and the reader is referred to his paper, which alludes to certain tacit postulates underlying this concept. It is important to distinguish experimental variables from variables which appear in theories, as it has sometimes been found that a measured variable (quite apart from being enumerable, which in many cases does not apply to theoretical variables) is not related to the theoretical variable in the way the theoretician believes. Thus, values of an electric current observed by means of a tangent galvanometer or other instrument are values of an *experimental* variable, while a current which appears in the mathematical theory of electrodynamics is a *theoretical* variable, and the hypothetical relationship of these two types of variables is a theoretical issue (see § 2.4A).

E. Statistical considerations

Popper (1967, p. 19) remarked in an essay on quantum mechanics that a statistical 'sample space [characterized by a statistical measure function] has hardly anything to do with the elements [of the population considered]'. Put differently: a statistical distribution function for a property whose observed samples form a class is not itself a property of the system. It is for this reason important to distinguish between a publicly observable property of a system and the statistical distribution of different samples of this property with respect to some parameter.

In *uncontrolled* public observations all variables are usually on the same non-manipulated footing, but in a controlled experiment it is convenient if possible to distinguish between independent variables which are under the manipulative control of the experimenter and *dependent* variables which change automatically as a result of changes of the independent ones, but in a fashion which cannot be rigorously controlled by the experimenter. For example, in many behavioural experiments stimulus variables are independent, and response variables dependent. Hebb (1954, p. 406) emphasised that in behavioural science the relations between the stimulus and response variables are often not of a regular nature. He wrote:

> I hope I do not shock biological scientists by saying that one feature of the phylogenetic development is an increasing evidence of what is known in some circles as free will; in my student days also referred to in the Harvard Law, which asserts that any well-trained experimental animal, on controlled stimulation will do as he damned well pleases. A more scholarly formulation is that the higher animal is less stimulus-bound (Goldstein, 1940). Brain action

> is less fully controlled by afferent input, behaviour therefore less predictable from the situation in which the animal is put. A greater role of ideational activity is recognizable in the animal's ability to 'hold' a variety of stimulations for some time before acting on them (Hebb and Bindra, 1952) . . . . There is more autonomous activity in the higher brain, and more selectivity as to *which* afferent activity will be integrated with the 'stream of thought', the dominant, ongoing activity in control of behaviour.

In some behaviour experiments, for example in 'operant behaviour' (Skinner, 1953, p. 65), stimulus variables cannot be explicitly defined, and only responses to a given situation can be observed.

If x denotes an independent (that is, manipulable) variable, then the experimenter can, subject to the limits of experimentally possible accuracy of measurement, allocate to x prescribed values

$$(x_i;\text{ for } i = 1, 2 \ldots n) \tag{2.3.1}$$

on different occasions of performing the experiment. If the variable is a numerical one, then x_i will be specified by a rational number, since a measurement can only be carried to a finite number of decimal points. The measured value of x_i is stated together with the statistically expected maximum error of measurement (such as 2.83 ± 0.003 cm). *The statistical error theory used is itself based on some important hypotheses.* If y denotes a dependent variable, then for each experimentally fixed value of x, say x_i, repetition of the experiment may yield a slightly, or perhaps considerably, different value of y (subject to a stated error in the measurement of y).

The differences in measured values of the dependent variable y (for the same observed approximate value of x) during repeat measurements of the same or a similar system are believed to be due to at least one of two major sources of error (see the discussion by Hogben, 1957, p. 214). (1) There may exist *constant* and *systematic* errors which point to faults in the instruments or experimental techniques used (for example, to faulty preparation of the observed system). (2) There always exist *random* errors. These could be due to (a) uncontrollable random fluctuations of internal parameters in a case where the *same* system is being repeatedly observed, or (b) slight differences in the systems concerned, in a case where similarly prepared systems are observed in repeat experiments. To illustrate case (2a): the temperature of a system might fluctuate excessively in some locality for a short time, thereby causing a local random change. Again, a preceding experiment with the same system could have slightly, but randomly, altered the internal structure or composition in some region of the system.

In case (2b), repetition of an experiment with a similarly prepared system could yield different results for y, because of random differences between the two systems. For instance, when studying the conductivity of wires, each of similar specified length, diameter and composition, different samples might contain differently distributed impurities, or vary minutely and uncontrollably in cross-section in certain localities. Errors of this kind cannot be eliminated but only minimised as far as possible. (For a general account of the theory of measurements, see Jánossy (1965).)

Suppose now that during the rth repetition of an experiment, with the same or a similarly prepared system, there corresponds to the same observed approximate value $\bar{x}_i$ of the independent variable a value $y_{i,r}$ of the dependent variable. For fixed value of i, $y_{i,r}$ may then be considered as a 'random variable'. The total interval over which y_i varies can be subdivided into a large number, say t, of distinct equally sized small intervals, and the measured value $y_{i,r}$ can then be allocated to the interval into which it falls. Let $m_{i,s}$ denote an *assumed* mean value of y_i for the sth interval. (One could, for example, hypothesise that y_i is normally distributed over the sth interval.) Suppose further that $f_{i,s}$ is the observed frequency of y_i values which fall into the sth interval in k repeat observations. Then for specified observed mean value $\bar{x}_i$ (subject to stated experimental error), the mean value $\bar{y}_i$, given by

$$\bar{y}_i = \left(\sum_{s=1}^{t} f_{i,s} m_{i,s}\right) \Big/ k \tag{2.3.2}$$

can be computed. The repeat experiments can be declared *consistent* if the values

$$\sigma_i^2 = \left[\sum_{s=1}^{t} f_{i,s}(m_{i,s} - \bar{y}_i)^2\right] \Big/ k \qquad (\text{for } i = 1, 2 \ldots n) \tag{2.3.3}$$

do not fall outside conventional (socially agreed) limits.

When dealing with non-numerical properties, for example, when a set of properties (or patterns) are assessed for mere presence or absence, we can evaluate the frequency $F_i(k)$ that property (or pattern) P_i turned up in k repetitions of an observation (that is, the frequency with which f_i assumed the value 1 in (2.2.3), p. 51). Genetics provides a rich source of such frequency statements (*cf.* Sinnot *et al.*, 1958). For example, in simple 'Mendelian' breeding experiments the relative frequencies of appearance of specific traits are being evaluated (and a 'trait' can be regarded as a 'property'). Quite generally, when observing $F_i(k)$, for a property P_i of any system, f_i may be postulated to be a random variable. Statistical 'significance tests' are then often used, on hand of the observed frequencies, to ascertain whether the frequency with which property P_i turned up is 'significant' (in an arbitrarily but intelligently defined sense) with reference to some specific hypothesis. For this purpose one has to discover whether, according to the significance test, the property turned up 'significantly' more frequently than could be expected if it appeared 'purely randomly'.[6]

[6] What one means by 'purely randomly' is either a matter of arbitrary definition, or depends on an indefinite number of conceivable empirical tests which could be taken as providing criteria of randomnes. As the number of conceivable tests need not be finite, any application of a finite set of tests for the randomness of a supposed 'random variable' is insufficient to ensure randomness. However, in some cases there exist escape routes from the customary interpretation of statistical significance tests (see Wassermann, 1955).

I must emphasise the following points. (i) As mentioned, all experiments are beset with experimental errors. In cases where numerical variables are measured, the permissible bounds of error are usually stated, subject to stipulated (that is, postulated) permissible maximal error deviations from mean values in the measurements of the independent and dependent variables (*cf.* Jánossy, 1965). Accordingly even so-called 'exact' sciences involve inherently statistical statements. In the case of non-numerical properties, where only the presence or absence of a property is being assessed, a human operator and his brain may serve as the detecting and evaluating system (for example, by inspecting electronmicrographs, or by describing the behaviour of animals of a particular species). When only the absence or presence of phenomena is being recorded it is meaningless to speak of a measure of experimental error. Claims could, however, then differ with regard to the frequencies of observation of phenomena. Experimenter A might claim that a particular strain of *E. coli* (bacterial) cells, when treated in a specified way, showed a certain property with a stated frequency, while experimenter B, on repeating A's experiment, may fail to confirm A's observations. Such cases are not exceptional. They are sometimes resolved by the discovery that experiments conducted in different laboratories have differed in subtle respects. For instance, the strains of bacteria used in one laboratory might have contained unsuspected types of mutants.

(ii) Criteria of acceptance of the consistency or inconsistency of results depend either on acceptance of relatively subtle statistical *conventions* (that is, 'significance tests'), or in other cases on blunt denials from two or more different laboratories. Such denials need not imply faulty experimentation. They may simply suggest that the experimental situation is more complicated than envisaged by either of the opposing camps.

(iii) The preceding considerations show that the celebrated cliché that 'numerical precision is the essence of science' (*cf.* Thompson, 1942; Needham, 1927; Lord Kelvin, cited by Cattell, 1950, p. 4; Hull *et al.*, 1940, p. 8; and Hull, 1951, p. 115) does not apply in many cases to properties themselves but only to statements relating to the *frequency* with which properties turn up. I stated that sometimes the properties themselves may be describable in terms of numerical variables. Yet in other cases this is not so. For instance, it is one thing to say that the human brain has (as a rule) a corpus callosum and that in some cases this may be congenitally absent, and to state the proportion of autopsies in a population which reveal a corpus callosum. But this is not the same as stating the estimated average number of nerve fibres of the corpus callosum of different subjects.

(iv) The amount of observational repetition required in order to establish *reliable* frequencies depends on the range of variation of the property considered and on the rarity of its occurrence. (Penrose (1957) illustrated by an interesting example, taken from genetics, the illegitimate use of a *single* observation in order to establish the equivalent of a frequency claim.)

(v) Statistical *limits of confidence* may decide the range of variation of a dependent variable which is to be accepted, although the choice of these limits is a matter of arbitrary convention (see above).

2.4 Empirical generalisations and inductions

A. Empirical generalisations[7]

I have argued that empirical statements describe publicly observable *properties* (see p. 60) of systems by means which suit the systems concerned or the state of development of the science to which the observations belong. In addition, empirical statements establish the frequency of occurrence of properties in repeat observations. As properties and the frequencies of their occurrence refer always to a finite set of samples of observed systems, theorists felt tempted to extrapolate (generalise) empirical statements from a *finite* set of observed samples to a hypothesised infinite population of samples of systems of the type examined. Mendel's laws, Ohm's laws, Boyle's law (and so on) represent such extrapolations, which are often referred to as *empirical generalisations*. For instance, Ohm's law is assumed to be satisfied by *all* wires of the same composition, length and cross-section (to a required degree of approximation). Some people, notably behavioural scientists, are inclined to refer to one or more unrelated empirical generalisations as a 'theory'. But I shall argue later that this practice is misleading, as empirical generalisations *per se* are not typical of properly constructed theories. They are only 'theoretical' in the sense that they extrapolate from immediate observations.

Empirical generalisations can take several forms. Where properties of systems can be specified in terms of *experimental variables* the empirical generalisation may replace (a) the average values (2.3.2) (which are computed for a finite set of repeat observations), and (b) the observed independent variables, by continuous variables. This leads then to a hypothesised relation between a theoretical 'independent' continuous variable $\bar{x}$ and the 'dependent' theoretical continuous variable $\bar{y}$, which is *assumed* to 'correspond' approximately (in a sense specified below; see pp. 71–72) to relations

[7] Some writers distinguish between 'laws' and empirical generalisations (e.g. Bunge, 1967c, vol. 1, § 6). Bunge calls an appropriate statement an 'empirical generalisation' provided that it cannot be deduced from an existing theory, that is, if it is an *isolated* generalisation. In contrast to this he calls an empirical generalisation a law if it can be deduced from the hypotheses of a theory. However, Boyle's and Ohm's laws were known before their theoretical deductions, from the kinetic theory of gases and the quantum theory of metallic conduction respectively. Again, various 'laws' for the frequencies of the spectral lines of the hydrogen spectrum were known long before their deduction from quantum mechanics (*cf.* Born, 1948, p. 100). I therefore see no reasons for distinguishing between empirical laws and empirical generalisations, since an empirical generalisation of today could, on Bunge's definition, become an empirical law to-morrow. The word 'law' also tends to conceal its hypothetical nature.

which would be obtained between $\bar{y}_i$ and $\bar{x}_i$ if the experiments were repeated indefinitely, and if $\bar{x}_i$ were replaced by a continuous variable which, within a certain range of values, could take *any* value. Hence it is assumed that experiments would satisfy the empirical 'law' approximately even for $\bar{x}$ values for which it has never been tested. For instance, the current and voltage variables which appear in Ohm's law are continuous variables, while the observed values in any experiment which tests Ohm's law (and mean values calculated from the latter) are discrete sets of values.

Correspondingly, frequency statements are generalised into probabilities. This is done by hypothesising that, if an experiment were to be continued indefinitely, then the numbers $f_{i,s}/k$ in (2.3.2) would tend to a unique limit $p_{i,s}$ which is called a probability (*cf.* von Wright, 1957, chapter 6). This interpretation of a probability as a hypothetical limit was first proposed by Venn, later adopted and axiomatised by von Mises (1919, 1931) and Reichenbach (1949) (for more recent axiomatizations, see Popper, 1963). A statistics based on the axioms of von Mises, or any related set of axioms, may be perfectly in order as a piece of formal logic or mathematics. However, when a statistics based on mathematical axioms is applied to empirical events, no *finite* set of observations could logically (in the sense of a rigorous mathematical proof) provide adequate evidence that applications of this mathematical system, to situations which implicitly generalise from observed frequencies to probabilities, are justified. I am not denying that statistics can be based on rigorous logical foundations, but the point at issue concerns its applicability.

Reichenbach (1949, 1961) attempted to make the use of probabilities more plausible by *interpreting* their applications as *betting ratios* or 'posits' which allow the user to make wagers about the future outcome of experiments or events. These posits need not be applicable for all time, and may stand in constant need of revision. In fact the probabilities posited by insurance companies for certain contingencies may change with social circumstances (such as the probability of death from certain diseases, which an actuary may 'posit' may have to be revised if new drugs become available). While the 'posit' interpretation of probability has heuristic value, and is appropriate to the trial-and-error attitude of science (see also Hogben, 1957; and see Salmon, 1966, for views related to those of Reichenbach), it does not *per se* make the jump from finite observed frequencies to corresponding probabilities any less arbitrary.

The generalisation from frequencies to probabilities may be interpreted as a special case of the human brain's capacity to form *hypotheses*. The important part which is likely to be played by brain-generated hypotheses in the solution of general problems was emphasised by Miller *et al.* (1960, p. 163). Probably generalisation by brains is mediated by mechanisms of the type used in concept and hypothesis formation.

To say that certain statistical distribution functions based on probabilities provide for the time being 'good bets' for the continuing behaviour of systems, and that we must alter these posits if experience forces us to do so, makes, as various people have noted, the application of statistics no less of a hypothetical enterprise. Even *the fact* that statistics *appears to work well in many theories and applications* (such as in quantum mechanics or statistical mechanics) provides no guarantee that there could not be 'erratic' natural phenomena for which it does not work. This does not mean that any obscure phenomena which can be backed by statistically designed experiments, and which cannot be explained in terms of contemporary theories, must be *ipso facto* interpreted as *prima facie* evidence for the inapplicability of statistics to certain natural phenomena. While the latter possibility can never be ruled out on grounds of logic, it always remains possible that such 'obscure phenomena' may become explicable in terms of new theories which are established in the light of new discoveries in well established branches of science (*cf.* G. S. Brown, 1957; Koestler, 1971; and Wassermann, 1955, for differing points of view).

Critical expositions of the probability concept were given by Hogben (1957), Braithwaite (1953), von Wright (1957), Reichenbach (1949) and Körner (1966), among many others. The use of probabilities to provide a measure for the 'degree of confirmation' of empirical generalisations and other hypotheses has been envisaged by various workers, notably by Carnap (1950, 1956). Körner (1966, chapter 9) has given an important critical discussion of Carnap's concept, which he considers for good reasons to be untenable. (The reader is particularly referred to Hogben's (1957) extensive critical discussion of the foundations of statistical theory, and to Körner's (1966) interesting remarks on statistics.)

Let us now consider an empirical generalisation of mathematical statements of type (2.3.2). Suppose that in a particular empirical generalization $\bar{x}$ is the only independent variable and $\bar{y}$ is the dependent variable. To extrapolate from the empirical data to a hypothesised 'mathematical law' one may assume, for example, as is often done, that the empirical data can be approximated to by a least-square curve fitting procedure. For this purpose we may assume some *arbitrary* functional relation of the form

$$\bar{y} = f(a_1, a_2 \ldots a_n, \bar{x}) \qquad (2.4.1)$$

where $a_1, a_2 \ldots a_n$ are a set of constants to be determined, and f is a postulated function (for example—*cf.* Weatherburn, 1952, p. 100—a polynomical of the form

$$f = \sum_{r=1}^{n} a_r \bar{x}^{r-1} \qquad (2.4.2)$$

where n is an arbitrarily chosen integer, or f could be represented by a finite number of terms of a Fourier series, etc.) We can then define the 'error' E_r

which arises for the empirical mean values of the variables $\bar{x}_i$, $\bar{y}_i$, for $i = 1$, $2 \ldots n$, by means of

$$E_i = f(a_1, a_2 \ldots a_n, \bar{x}_i) - \bar{y}_i \tag{2.4.3}$$

Following this we can determine the coefficients a_i (for $i = 1, 2 \ldots n$) so that $\sum_{r=1}^{n} E_r^2$ becomes a *minimum*, which requires as one set of conditions that $\partial f / \partial a_i = 0$, for $i = 1, 2 \ldots n$. From these conditions one can often, with the help of computational techniques, obtain the a_i.

It must be emphasised that the least-square approximation procedure presents only one *fashionable* method for obtaining *arbitrary* interpolation functions which approximate to empirical mean values. It can be shown that the least-square method gives a 'best possible fit' according to *particular criteria* of 'goodness of fit' (see Hogben, 1957, p. 200). But as we could chose different criteria, we cannot on *a priori* grounds justify the superiority of any criterion as compared with any others which may be conceivable.[8] Any criteria are arbitrary, and if widely adopted (such as the least-square method) represent, at best, social conventions among scientists.[9]

Any claim to have a 'best fitting' curve for a set of mean values can only refer to a whole set of arbitrary procedures, and is therefore open to complete scepticism, and may be in need of subsequent revision. Some writers thought it possible to discover *systematic* methods for deriving best possible data extrapolation procedures according to postulated 'degrees of simplicity' (*cf.* Jeffreys, 1957; Popper, 1960a; Kemeny, 1953a). More generally, the recurring idea that a *systematic theory of induction* can be constructed is open to grave doubts and objections (see Ackermann, 1961; Hogben, 1957, pp.

[8] Bunge (1967c, part 1, p. 321) believes that wherever possible, one should base mathematical laws on 'exact functions' (such as sines and cosines), instead of using polynomical approximations. However, although the *definitions* of these functions are exact, their evaluations also involve approximations based on infinite series which are broken off after a finite number of terms. Bunge's suggestion that exact functions are to be preferred for empirical generalisations, as they can sometimes be deduced from higher order theories, is useful but restricted in practice. Such simple final terms do not appear in the approximation procedures used in many really powerful theories (for example, in self-consistent field theories or in molecular orbital theory in quantum chemistry).

[9] The principle of 'economy of hypotheses' is sometimes appealed to. The principle *per se* is also arbitrary, as it is not self-evident that natural phenomena must conform to any economy of hypotheses, in a case where there exists a choice of alternative explanations, and the statement of the 'principle' is often misleading. Economising in hypotheses must not be confused with 'Occam's razor', which states that postulated *entities* (such as elementary particles) should be as few as possible. Entities are not the same as hypotheses, as each postulated scientific entity must be *defined* in terms of hypothesised properties. Occam's razor therefore does not imply a necessary economy in the number of hypotheses used, while the principle of economy of hypotheses does. However, in theories, where hypotheses are often complexly interlinked *via* a hypothetico-deductive network (see below) it may be difficult to determine which hypotheses are completely independent of others, particularly if a theory is based on a combination of other theories. Sometimes for clarity people may formalise a 'mechanism' by dividing its description into numerous subhypotheses, which other theorists would lump together into a single hypothesis.

25 and 324). (The reader is referred to Feigl (1956, pp. 31–5) and Carnap (1950) for different points of view.) As an extrapolation always goes beyond facts, *it must remain a hypothesis*, and no hypothesis referring to an extrapolated infinite class can be substantiated by a finite set of members of that class. This does not mean that existing procedures have no pragmatic value (*cf.* Kraft, 1966, p. 310). In fact, the ability of human brains to guess plausible hypotheses is substantial. Evidence for this ability is documented at many levels of theorising. If we consider Einstein's general theory of relativity, Dirac's theory of the spinning electron and his formulation of quantum electrodynamics, Pauli's exclusion principle, Schrödinger's and Heisenberg's quantum hypotheses, Bohr's early theory of the atom, Maxwell's electrodynamics and, above all, Newton's formulations of mechanics, we see that human brains have a remarkable capacity for making inspired (non-systematic) guesses even when it comes to the formulation of very subtle hypotheses.

Returning to least-square fits, I must emphasise that the fitted functions will not necessarily be satisfied by any of the mean values $\bar{x}_i$, $\bar{y}_i$, for $i = 1, 2 \ldots n$ which were used in establishing the fit. The fitted functions simply represent an 'idealised' approximation. In order to *refute* such an idealised empirical generalisation experimentally, one requires mean values $\bar{y}_i$ for given values $\bar{x}_i$ for a considerable number of values of i, and has to show that these values of $\bar{y}_i$ lead to errors, given by (2.4.3), which exceed in magnitude certain conventionally (but nevertheless arbitrarily) agreed limits.

B. Behaviour rules and 'pretence explanations'

I have already mentioned that our everyday languages contain implicit primitive interactionist hypotheses (pp. 28–29). In addition they make use of expressions which introduce 'pretence' causal explanations of phenomena. Closer inspection of such 'pretence explanations' shows that they do not operate in terms of postulated mechanisms or processes. At best they 'explain' a state of affairs x by implying tacitly, on an *ad hoc* basis, that x forms a special case of a simple verbal empirical generalisation. For example, if I say 'John eats, *because* he is hungry', then the word 'because' does not relate John's eating behaviour, *via* brain mechanisms, *causally* to his previous speech behaviour, for example to an announcement he may have made that he is hungry. Despite its deceptive causal appearance the word 'because' functions in this statement, and innumerable similar statements of everyday language, simply as an *ad hoc* hypothesis. In the present case it implies that John conforms to the empirical generalisation or 'behaviour rule' that 'people who say they are hungry are likely to eat when confronted with food'. (See also Skinner (1953, p. 31) and Golightly (1952) for related but different points of view.)

Many 'behaviour rules' concerning human or animal behaviour are non-numerical empirical generalisations. Some philosophers have claimed that such 'behaviour rules' should suffice to *explain* human behaviour in terms of a 'rule following model' (Peters, 1958). However, as behaviour rules are empirical generalisations which by themselves do not provide explanations in the sense of theories (see § 3.3 below), behaviour rules do not genuinely 'explain' anything. If we observe that a particular gas conforms to van der Waals' equation of state in good approximation, this does not *explain* how molecular mechanisms produce that behaviour of the gas, which is approximately expressed by its equation of state. Only a theory, namely the kinetic theory of gases, can provide the required explanation.

A hypothetico-deductive theory serves to *explain* empirical generalisations by deducing them from postulated initial hypotheses (see below) and by interrelating (often only approximately) a plurality of empirical generalisations indirectly *via* initial hypotheses and secondary hypotheses which are deduced from the initial ones. An 'isolated' empirical generalisation, which is unrelated to others *via* a theory, can only provide 'explanations' in the sense of subsuming particular properties (or particular average values in the case of measurable variables) as special cases, which approximately fit its general statements. We say then that the properties (and so on) *conform* to the empirical generalisation.

C. Induction

In science the word 'induction' refers to those procedures which are being applied for obtaining generalisations; that is, they include verbal generalisations,[10] least-square fittings, and so on. Hume, in his *Treatise on Human Nature* (1739/40), argued convincingly that the procedures of inductive generalisation cannot be justified by any rational arguments. They are inherently open to scepticism and their validity cannot be proved. This conclusion remains unshaken, notwithstanding numerous arbitrary attempts to 'justify' induction. Improved neuropsychological theories should be able to explain how brains can generalise. However, this alone would never be adequate to show that this brain capacity is invariably being correctly applied; that is, it could not 'justify' the inevitable validity of *particular* generalisation procedures.

In addition there are empirical problems. Popper (e.g. 1960a) has repeatedly emphasised that scientific hypotheses are only valid until they become decisively falsified empirically. Accordingly the validity of any induction can only be upheld while it is being further confirmed, or in the absence of negative evidence. This does not mean that some empirical generalisations could not be indefinitely valid, or at least for historical periods which are very long compared to the human life span. But we can never be sure of this. For all we know (and such suggestions have been made repeatedly) certain 'constants' of nature might be changing slowly with time

[10] See also Popper's (1960) interesting historical paper.

(*cf.* Anon, *Nature*, 1971, **234,** 505). Accordingly, beliefs that any empirical laws (and likewise higher order hypotheses of theories; see below) are universally valid for all time are unjustifiable bits of metaphysics. Such beliefs rest on the delusion that studies on this planet covering the relatively short time span of organised scientific work (say 300 years) have revealed some laws which are bound to hold for an eternity. While this could well be the case, we shall never know.

Von Wright (1957), whose scholarly book should be consulted, carefully surveyed numerous futile attempts to justify induction (see also Kemeny's (1953) review). The situation is even more complex. Feigl (1964, p. 49) stressed that hypotheses can be knocked out but subsequently reinstated. Some cases of such 'ups and downs' are known. For instance, the debates centring round the 'steady-state' and 'big bang' theories of cosmology present a typical example of the fluctuating status of a theory (*cf. Nature*, 1973, **246,** 378, and Tryon, 1973).

Many people have argued that they see no need for a logical justification of *any* procedure which *works*, and that pragmatism is the best guide. They say that sceptics should certainly act as watchdogs, but scepticism should not bar us from making experiments and formulating theories. If empirical generalisations and theories provide a better understanding and linkage of phenomena, and in addition permit numerous predictions, then we are certainly better off with theories than without them, irrespective of whether we can provide 'justifications' for the methods of generalisation. Perhaps some sceptics feel that unless they can logically 'justify' a particular choice of methods for obtaining empirical generalisations or hypotheses of theories, they are no better off than a believer in orthodox religion, who also cannot justify his beliefs. However, I suspect that few people would dispute that despite their ultimate lack of 'logical justification' the insights offered by electromagnetic theory, quantum mechanics, molecular biology, and other major scientific achievements, are different from metaphysical religous pronouncements. The fact that two different realms of belief are beset with uncertainties does not mean that their uncertain insights are of comparable kinds. The irremovable *uncertainties* of the foundations of science can only give displeasure to people who look upon science as a substitute for religion, in the misguided hope that science, in contrast to religion, offers doubtless certainty. In reality both scientific and religious people are believers, though of a different kind.

2.5 Conventionalism and other misconceptions

That most, if not all, empirical generalisations are explicitly or implicitly of a statistical nature is not as widely appreciated as it ought to be. The fact that empirical generalisations must be based on *repeat* observations of properties suggests already that we are dealing with a statistical phenomenon, and some of the discussions of § 2.4 aimed to bring home this point. Even when a human observer only inspects a system for repeated presence or absence of a property his brain may act as the decision making or 'sampling' device. Reichenbach (1961) and others recognised clearly the essentially statistical basis of empirical generalisations, while some experimentalists failed to do so.

For example, G. B. Brown (1952) cited spurious correlations as evidence that 'anything can be proved by statistics'. Some defenders of this opinion believe that Ohm's and Boyle's laws are 'exact relations', and independent of statistical considerations. But *de facto* these, like other 'exact laws', depend on repeat observations and averagings together with hypothetical extrapolations. Potentially any *isolated* empirical generalisation, however 'exact', could represent a spurious correlation (or could be based on an unsuspected statistical bias). It is only when *several* empirical generalisations become meaningfully linked within a theory (see below) that we can decide (but never be certain) that these empirical generalisations are not spurious. The frequency with which an empirical generalisation can be confirmed is not necessarily an indicator of its non-spuriousness, as there is no reason why a spurious correlation might not turn up with high frequency. On the other hand, difficulties experienced in repeating observations of relatively rare phenomena such as major earthquakes or particular volcanic eruptions do not mean that the phenomena are spurious.

As empirical generalisations cannot be shown to be unique, because they rely on logically unjustifiable procedures, *conventionalists* were at pains to stress this use of 'arbitrary conventions' (least-square fits, and so on). However, the situation is not quite as tragic as they try to make out. Although 'best fit' methods are unquestionably arbitrary, their continued use over decades has proved their value not only in prediction but also in confirmations of experiments. Hence to elevate logical uncertainty into a melodrama of 'the vice of arbitrariness' is as unjustifiable as the fitting methods themselves. As long as one does not make the error of assuming that empirical laws are 'eternal truths', serious danger can be averted.

Some conventionalists thought that empirical generalisations are equivalent to definitions. For example, Kapp (1955, p. 59) argued with conviction that Ohm's law *only* defines electric resistance. However, with the given *operational definitions* of current and voltage measurements, it is hardly a convention that the extrapolated relation (based on curve fitting) between the applied voltage and current in a wire should (with the adopted fitting method) regularly turn out to be approximately linear (for restricted voltage ranges). The linearity of the relationship is not exactly a matter of convention, since the same method of fitting could also have led to a non-linear relationship. In fact, for strong electric fields (for example, in dielectrics near the breakdown voltage) the relation is found to be non-linear (*cf.* Whitehead, 1951, p. 101).

Further arguments against the narrower forms of conventionalism were given by P. Frank (1957, pp. 6 and 76) and Schlick (1948, chapter 3), among others.

More generally, conventionalists even suggested that initial hypotheses of theories (see below) are disguised definitions (*cf.* Poincaré, 1905, chapter 6; Duhem, 1954, pp. 280ff; who were cited by Bunge, 1967c, vol. 1, p. 138). Bunge stressed that conventionalists are apt to regard all hypotheses, including *ad hoc* ones, on the same footing. In addition they are 'instrumentalists' who consider scientific theories as deductive (and predictive) instruments which, when mathematical formulations are used, should employ all symbols *purely formally*, and that semantic interpretations of higher order 'unobservables' are unessential (*cf.* Bunge's (1967c, vol. 1, p. 488) remarks concerning the conventionalist view of potentials in field theory, Schrödinger functions, and so on).

Poincaré's (1898) conventionalist thesis, cited by Čapek (1969, p. 410) asserts that

> . . . Our choice [of hypotheses] is therefore not imposed by experience. It is simply guided by experience. But it remains free; we choose this geometry rather than that geometry, not because it is more *true*, but because it is more *convenient*. To ask whether the geometry of Euclid is true and that of Lobatchevski is false, is as absurd as to ask whether the metric system is true and that of the yard, foot, and inch is false.

Poincaré's 'convenience postulate' disregards the fact that in practice it is hard enough to discover *one* set of scientific hypotheses which fit satisfactorily a *wide range* of phenomena. Multiple alternatives are almost entirely 'conventionalist nightmares' not found in practice. For example, the older quantum mechanics (based on the Bohr–Sommerfeld–Wilson theory) enabled Sommerfeld in 1916 to derive a formula for the discrete energy levels of the hydrogen atom, which is *formally* identical with that obtained by Dirac's later relativistic theory, both formulae being deduced from totally different hypotheses (see Born, 1948, p. 159). Amazing as this coincidence may be, it does not strengthen the conventionalist case, since the later quantum mechanics could account more adequately for many data than the older quantum theory, and explain a host of phenomena which the latter could not explain at all.

2.6 Formal logic and statistics in the formulation of empirical generalisations

My approach to empirical generalisations differs from that of Körner (1966, chapter 5). His book has, among other features, the great merit of having drawn the attention of philosophers to the inexactness of empirical classes (that is, the considerable variation of members of classes of observed things and events). While this is well known to scientists, this variability appears to have been disregarded by some philosophers of science (*cf.* Körner's (1966, p. 89) remarks). I suggested earlier (see p. 48) that human brains can express class membership of a configuration, by means of token representations, and that the latter are the physical representatives of concepts. Moreover, the brain has apparently the ability to allocate many variants of a pattern to the same concept (see the passage from Brain on p. 18).

Brains may allocate objects directly to classes (as happens, for example, in many anatomical studies). Alternatively, they may use measurable or other indirectly observable properties of systems, and classify these. In either case the classification proceeds as a result of the capacity of brains to allocate a range of variants to a class. This allocation can, as stated, be direct, or it can be the result of a brain-invented class-allocation procedure (such as the curve fitting procedures discussed earlier). In all cases considered, statistics may be used either indirectly (as in curve fitting) or directly (for example, when an anatomist states the observed frequency of a particular property among an observed set of samples, or a clinician cites the frequency of successes of a drug). I remind the reader that such considerations involve the axiomatic (logical) foundation of statistics (*cf.* Popper, 1963) together with logically unjustifiable decisions concerning the applicability of statistics to empirical situations. However, the axiomatics, as well as the decision to apply statistics, are both brain generated.

Statistics, either explicitly, or implicitly, takes care of the 'inexactness' of empirical classes, without our having to introduce a special logic to cover the case (apart from the logical axiomatic foundations of statistics itself). I see no need to describe scientific procedures by means other than those which are actually used, and to describe what is typical about them. But the typical evaluative features of scientific observations can be expressed in terms of statistics (together with arbitrary decisions), without the aid of a specially invented formal logic.

Statistics is not only used in such disciplines as quantum mechanics and statistical mechanics but, as we saw, enters implicitly or explicitly into the classification of every type of observation. In some cases the implicit use of statistics is less obvious. Take, for instance, Körner's (e.g. 1966, p. 76) repeatedly cited empirical generalisation that 'all magnetised pieces of iron attract iron filings'. This is on the same level of generalisation as the statements (a) 'all metals of a specific kind conduct electricity', (b) 'all gases have a temperature above the absolute zero temperature'. Such low level generalisations are equivalent to statements of the type 'this system has property x with probability 1', where property x can vary over a wide range. In the case of magnetised iron the 'property' refers to the attraction of iron filings (when sufficiently close to the magnet), while in cases (a) and (b) the properties are the existence of a measurable current and of a measurable temperature respectively.

However, these generalisations are far less sophisticated than the equations of state of gases or Ohm's law, which assert generalised *relations* between mean value-representing variables. In the case of magnetised iron, a more exact generalisation would require specification of mathematical equations for the pattern of arrangement of iron filings round a bar magnet of postulated shape and dipole moment. (Elementary textbooks usually proceed the other way round. They postulate a theory based on magnetic 'poles' located near either end of the bar magnet (such poles lack a proper theoretical basis since, according to electromagnetic theory, only electric currents within atoms or on a molar scale produce magnetic 'forces'). From the theory, the 'lines of force' are computed, and it is postulated that these determine the arrangement of the iron filings. Hence the equations *derived from the theory* may be regarded as the 'empirical generalisation' which has to be tested. In the preceding example, an empirical generalisation is not derived in the first instance by least-square fittings, but is generated by a theory. However, the mathematical equations of the lines of force may be regarded as being equivalent to an assumed empirical extrapolation function.)

On p. 67 I stated that with a property P_i of an observed system can be associated a frequency $F_i(k)$ of its appearance in k independent observations. This can in appropriate cases tentatively be generalised into a corresponding hypothesised probability $p(P_i)$ that property P_i should turn up in a sufficiently large number of observations. In fact, we can interpret the probability $p_{i,r}$ introduced on p. 70 as a special case of a general probability assignment to a property. The relevant 'property' is then a statement, namely that for a fixed mean value of the independent variable $\bar{x}_i$ (subject to stated error) the dependent variable y_i falls, when measured, into the sth interval, which is defined on p. 67.

More generally, one could have probabilities for the joint occurrence of several properties; for example, $p(P_{r_1}, P_{r_2} \ldots P_{r_m})$ is the probability for the joint occurrence of the properties $P_{r_1}, P_{r_2} \ldots P_{r_m}$, which form a subset of the total set of properties under consideration for a particular system or event. (I remind the reader that, according to p. 50, the suffixes r_i can in many cases be interpreted as a *set* of time-dependent systemic parameters.) In the most general case the values assumed by the time parameter in the systemic parameters which appear in (2.2.2) on p. 50 need not be the same for different r_i. This enables us to state the joint probability that the system has, say at time t_1 property P_{r_1}, at time t_2 property P_{r_2}, and so on, where r_1 is the appropriate set obtained from (2.2.1) and (2.2.2) when t is equated to t_1, and similar statements apply for r_2, and so on.

3 THE FORMAL STRUCTURE OF HYPOTHETICO-DEDUCTIVE THEORIES

3.1 Initial hypotheses: questions of choice

The structure of theories has been discussed by many writers (e.g. Braithwaite, 1953; Carnap, 1928, 1938a, b, 1946, 1950; Brody, 1970; Woodger, 1947; Hempel and Oppenheim, 1948; Bunge, 1967b, c). It is desirable to distinguish a theory's *reconstructed* logical structure from the historical order of discovery of its various parts. Although the history of science shows many irrational or dubious ideas which, while later eliminated or improved on, led ultimately to important theories, an appreciation of heuristic or 'intuitive' ideas which inspired a theory is often as valuable as the study of its logical reconstruction. Heuristic considerations of inventors of a theory sometimes provide deep insights into *motivations* which led to certain formulations. For instance, Bohr's 'correspondence principle' was motivated by the notion that quantum mechanics should resemble classical mechanics, wherever possible, in its formal features. Those modern texts which are based on the 'Schrödinger representation' of quantum mechanics obscure this fact (since Schrödinger functions or Schrödinger equations have no near equivalents in classical mechanics; but see Sommerfeld's (1939) treatment of the transition from the Hamilton–Jacobi equation to the Schrödinger equation). However, older treatments (*cf.* Born and Jordan, 1930; see particularly their preface) show that one can develop quantum mechanics without a Schrödinger representation, and according to the correspondence principle (see also Heisenberg, 1930, pp. 105ff), even if the Schrödinger representation is more useful in solving many practical problems than some other representations.

The example just cited exemplifies an important aspect of many highly developed theories. There often exists a *multiple choice of formulations of initial hypotheses* for such theories. For instance, for classical mechanics one may choose a modernised formulation of Newton's hypotheses as initial hypotheses. Alternatively, as is fashionable in some treatments (see, for instance, some recent textbooks), one could start from a variational principle of the form

$$\delta \int_{t_1}^{t_2} L \, \mathrm{d}t = 0 \qquad (3.1.1)$$

where L is Lagrange's function. This type of initial hypothesis suffers from a drawback. While for many (holonomic conservative) systems one has discovered general rules for constructing the Lagrange function, for many other systems one has to fall back first on Newtonian mechanics in order to construct the Lagrange function; that is to say, there exists no generally known procedure for constructing Lagrange functions *ab initio*.

For instance, the Lagrange function for a charged particle moving in an electromagnetic field (in Gaussian units) is of the form

$$L = T - (e/c) \sum_{s=1}^{3} (\dot{x}_s A_s) + e\varphi \tag{3.1.2}$$

where T is the kinetic energy of the particle, e its charge, c the vacuum velocity of light, x_s, for $s = 1, 2, 3$, the coordinates of the particle, and A_s, for $s = 1, 2, 3$, and φ the three components of the vector potential and the scalar potential of the field respectively. To obtain a Lagrangian of type (3.1.2) requires some mathematical manipulation, for the simple reason that electromagnetic force fields do not form an integral part of classical mechanics. But the example shows that (a) one *can* form an appropriate Lagrange function, and (b) the mode of its construction is far from obvious. (Moreover, the 'Lagrange density function' of field theory—*cf.* Wentzel, 1949 for the definition—is not unique, and different 'Lagrange density functions' can yield the same field equations; *cf.* Bunge, 1967b, p. 47).

When expounding quantum mechanics, one could, like Dirac (1958), start with certain hypotheses concerning quantum mechanical 'Poisson brackets'. Alternatively, following Pauli's (1933) celebrated article in the *Handbuch der Physik*, one could start with a Schrödinger representation (see also Kramers, 1938, for an approach along the same lines). (For an elegant derivation of the Schrödinger representation, starting from a matrix representation, see Heisenberg, 1930, pp. 132ff.) The preceding illustrations show that one can develop the same 'reconstructed' theory by starting from different hypotheses. For instance, one can *deduce* the Lagrangian formulation of mechanics from the Newtonian formulation and, *vice versa*, one can deduce much (but not all) of the Newtonian formulation from the Lagrangian theory. While it is interesting to expound systematically one possible 'comprehensive set of postulates' of a theory (for example, of quantum mechanics, *cf.* Bunge, 1967d), it would appear that there exists no *unique* way of axiomatising a theory. There appear to be several different, sometimes partly, sometimes completely equivalent ways to make axomatic representations of a theory.

3.2 Examples of initial and deduced hypotheses

A. Newtonian mechanics of a point particle

From now on I shall formally denote the *i*th initial hypothesis of a particular theory by H_i, so that a set of initial hypotheses of a theory can formally be denoted by

$$[H_i; \text{for } i = 1, 2 \ldots p] \tag{3.2.1}$$

According to § 3.1, the set (3.2.1) is not unique for many particular 'reconstructed' theories. I must also emphasise that the H_i of a properly constructed theory are *not* empirical generalisations.

As a first example, I shall list some (but not all) of the initial hypotheses of one version of the reconstructed Newtonian theory of point particles. (For a different approach, see Bunge, 1967b.)

H_1: Particles are located at points; that is, they lack extension and are structureless. A 'point particle' located at a point P can be assigned a position vector $\boldsymbol{r}$, which is defined[1] by a triplet of components (x_1, x_2, x_3), which are the Cartesian (length) coordinates of P (see appendix) referred to a right-handed rectangular Cartesian frame of reference. (It is interesting to note that other non-Newtonian 'classical' particle theories have abandoned the notion of a point particle, and have postulated extended particles—Born, 1934; Born and Infeld, 1934a, b, 1935.)

H_2: With each point particle there is associated a positive numerical quantity m called its 'intertial mass', which is a constant.

H_3: Each point particle is assigned a gravitational mass, assumed to be equal to its inertial mass (in suitable units).

H_4: A point particle moving relative to the right-handed rectangular Cartesian frame K, with respect to which its position vector $\boldsymbol{r}$ is defined, is said to be acted on by an inertial force $\boldsymbol{F}$, defined by

$$\boldsymbol{F} = m\,\mathrm{d}^2\boldsymbol{r}/\mathrm{d}t^2 \tag{3.2.2}$$

If referred to frame K, the particle has an acceleration $\mathrm{d}^2\boldsymbol{r}/\mathrm{d}t^2$ at time t. The time t is assumed to be measured in K by chronometers which each indicate the same quantity t, even if located at different points in K; that is, all chronometers in K give the same reading, defining therefore a 'simultaneous' time relative to K. A frame for which the definition (3.2.2) applies is called an '*inertial frame*' (*cf.* Joos, 1947, p. 217).

The definition (3.2.2) is often referred to as Newton's second 'law'. But, as (3.2.2) is a *definition* (of $\boldsymbol{F}$), and not an empirical generalisation, the word 'law' is a misnomer. However, (3.2.2), though a definition, is *also* a hypothesis, since it defines $\boldsymbol{F}$ in a specific mathematical form in relation to m and the acceleration. (In fact, the particular form of the postulated relation (3.2.2) enables one to define an 'inertial frame'.)

H_5: If a point particle is accelerated, its inertial force is assumed to be equal to a 'constitutive force' $\boldsymbol{F}$ exerted on the particle by other material systems. The nature of the constitutive force (that is, its mathematical form)

[1] For readers not familiar with vector analysis I have briefly provided a few essentials in the appendix at the end of this book.

depends specifically on the system which exerts it. (For instance, typical constitutive forces are produced by gravitational fields or electromagnetic fields. Other examples are elastic forces, tensions in strings acting on (idealized) 'point particles' attached to strings, and so on.)

H_6: For a *system* of point particles (P_i; for $i = 1, 2 \ldots n$), it is assumed that (referred to an inertial frame) particle P_i has mass m_i and position vector $\boldsymbol{r}_i$ in K, and that the constitutive force $\boldsymbol{F}_i$ acting on P_i is due to a vectorial superposition of two types of forces:

(i) First there may be 'external forces', exerted on P_i by systems which do not form part of the system of particles considered. I shall denote by $\boldsymbol{F}_i^{\text{ex}}$ the resultant external force acting on the ith particle. For instance, if according to Newtonian mechanics one considers the molecules in a container of gas (in approximation) as a system of point particles, then the gravitational field force produced by the earth, which acts (by hypothesis) on any gas molecule, could be considered as an 'external force' acting on that 'particle'. In some cases the external forces acting on P_i may depend on the velocity $\mathrm{d}\boldsymbol{r}_i/\mathrm{d}t$ of P_i. For instance, when an electromagnetic field acts on a system of charged particles, we regard, according to Newtonian mechanics, the electromagnetic field as providing the external forces acting on the particles. The external force $\boldsymbol{F}_i^{\text{ex}}$ acting on the ith particle can then be written in the form (the vectors $\nabla\varphi$ and curl $\boldsymbol{A}$ being defined in the appendix):

$$\boldsymbol{F}_i^{\text{ex}} = -e_i[-\nabla\varphi - (1/c)\,\partial\boldsymbol{A}/\partial t + (1/c)(\mathrm{d}\boldsymbol{r}_i/\mathrm{d}t) \wedge \operatorname{curl}\boldsymbol{A}] \qquad (3.2.3)$$

where $\boldsymbol{A}$ and φ are the values of the vector and scalar potential at the position of the ith particle, e_i its charge, c the vacuum velocity of light, and Gaussian units are being used. (Expression (3.2.3) is used in constructing the Lagrange function L given by (3.1.2) for a single particle.) I shall not discuss under which conditions (for example, in classical statistical mechanics) Newtonian mechanics could be applied to gas molecules or to certain types of charged particles. The preceding examples only serve to illustrate the concept of 'external field'.

(ii) There are assumed to be 'internal forces' which act pairwise between the particles of the system. These forces are postulated even if the system were to be 'completely isolated' (an idealisation which is in practice impossible). I shall denote by $\boldsymbol{F}_i^{\text{in}}$ the resultant 'internal force', due to all other particles, which acts on the ith particle. Newtonian mechanics assumes that, if the jth particle of a system exerts on the ith a force $\boldsymbol{f}_{ij}$, then

(a) $$\boldsymbol{f}_{ij} = -\boldsymbol{f}_{ji} \quad \text{and} \quad \boldsymbol{f}_{ii} = 0 \qquad (3.2.4)$$

(This is one way of formulating Newton's 'third law', which, however, is not an empirical generalisation but an initial hypothesis of Newtonian mechanics.)

(b) $\boldsymbol{f}_{ij}$ lies along the line joining P_i and P_j; that is, it is parallel to the vector $\boldsymbol{r}_i - \boldsymbol{r}_j$.

(c) The total internal force due to all other particles of the system acting on the ith particle is of the form

$$\boldsymbol{F}_i^{\text{in}} = \sum_{j=1}^{n} \boldsymbol{f}_{ij} \tag{3.2.5}$$

Hence the total force acting on the ith particle due to external systems and other particles of the system to which the ith particle belongs is assumed to be of the form

$$\boldsymbol{F}_i = \boldsymbol{F}_i^{\text{ex}} + \boldsymbol{F}_i^{\text{in}} \tag{3.2.6}$$

where $\boldsymbol{F}_i^{\text{in}}$ is given by (3.2.5).

(d) The equations of motion for the system of particles are then, according to (3.2.2), (3.2.5) and (3.2.6), assumed to be of the forms

$$\boldsymbol{F}_i^{\text{ex}} + \sum_{j=1}^{n} \boldsymbol{f}_{ij} = m_i \, \mathrm{d}^2\boldsymbol{r}_i/\mathrm{d}t^2 \qquad (\text{for } i = 1, 2 \ldots n) \tag{3.2.7}$$

H_7: Initial conditions (such as the positions and velocities of the particles at time $t = 0$) have to be specified.

It is seen that the preceding partial formalisation of classical (Newtonian) mechanics makes use of several *calculi*. Among others, it uses *vector analysis* (which is an algebra of ordered number triplets; *cf.* Jeffreys, 1961, and the appendix) and differential and integral calculus. The usual strategy (for a single particle) is to combine (3.2.2) with the constitutive definition of force, and to eliminate the forces thus defined, thereby obtaining a differential equation for the motion of the particle, which can then in many cases be solved either formally or by means of computers, subject to given initial conditions. Thus, vector analysis and differential and integral calculus, ordinary algebra (and so on) are used as 'deduction calculi' in deriving deductions from the initial hypotheses of classical mechanics.

A few examples may help to illustrate how, starting from the initial hypotheses of classical mechanics, we can derive secondary hypotheses (deduced hypotheses) which *per se* are not empirical generalisations, but which are useful in deriving empirical generalisations. Additional examples will show that in some cases one can directly derive presumptive empirical generalisations from the initial hypotheses.

1. For a system of particles one *defines* the position vector $\boldsymbol{g}$ of the centre of gravity by

$$\boldsymbol{g} = \sum_{i=1}^{n} m_i\boldsymbol{r}_i/M, \qquad \text{where} \qquad M = \sum_{i=1}^{n} m_i \tag{3.2.8}$$

Summing (3.2.7) over the suffixes i for all particles of the system, noticing that on account of (3.2.4)

$$\sum_{i=1}^{n} \sum_{j=1}^{n} \boldsymbol{f}_{ij} = 0 \tag{3.2.9}$$

and using (3.2.8), we obtain

$$\sum_{i=1}^{n} \boldsymbol{F}_i^{\text{ex}} = \sum_{i=1}^{n} m_i \, \mathrm{d}^2\boldsymbol{r}_i/\mathrm{d}t^2 = M \, \mathrm{d}^2\boldsymbol{g}/\mathrm{d}t^2 \tag{3.2.10}$$

Equation (3.2.10) represents an important *deduced hypothesis*. In words it states that the vector sum of the external forces acting on a system of particles equals the acceleration of a mass equal to that of the total system placed at and moving with the centre of gravity of the system.

2. One *defines* the angular momentum $\boldsymbol{A}_{\mathrm{O}}$ of the system of particles about an origin O of an inertial frame [see the appendix, formula (A.8), for a definition of the symbol $\wedge$] by means of

$$\boldsymbol{A}_{\mathrm{O}} = \sum_{i=1}^{n} \boldsymbol{r}_i \wedge (m_i \, \mathrm{d}\boldsymbol{r}_i/\mathrm{d}t) \tag{3.2.11}$$

Also from (3.2.7) we obtain that

$$\sum_{i=1}^{n} \boldsymbol{r}_i \wedge \boldsymbol{F}_i^{\text{ex}} + \sum_{i=1}^{n} \sum_{j=1}^{n} \boldsymbol{r}_i \wedge \boldsymbol{f}_{ij} = \sum_{i=1}^{n} \boldsymbol{r}_i \wedge (m_i \, \mathrm{d}^2\boldsymbol{r}_i/\mathrm{d}t^2) \tag{3.2.12}$$

(where $\boldsymbol{a} \wedge \boldsymbol{b}$ denotes a vector product). But the double sum on the left contains the terms

$$\boldsymbol{r}_i \wedge \boldsymbol{f}_{ij} + \boldsymbol{r}_j \wedge \boldsymbol{f}_{ji} = (\boldsymbol{r}_i - \boldsymbol{r}_j) \wedge \boldsymbol{f}_{ij} \tag{3.2.13}$$

(as follows from (3.2.4)). Since, according to H_6 (ii, b), $\boldsymbol{f}_{ij}$ is parallel to $\boldsymbol{r}_i - \boldsymbol{r}_j$, it follows that the right-hand side of (3.2.13), and hence the double sum in (3.2.12), vanishes. Also from (3.2.11) we see that

$$\mathrm{d}\boldsymbol{A}_{\mathrm{O}}/\mathrm{d}t = \sum_{i=1}^{n} \boldsymbol{r}_i \wedge (m_i \, \mathrm{d}^2\boldsymbol{r}_i/\mathrm{d}t^2) \tag{3.2.14}$$

Combining (3.2.12) with (3.2.14) we obtain

$$\boldsymbol{M}_{\mathrm{O}} = \mathrm{d}\boldsymbol{A}_{\mathrm{O}}/\mathrm{d}t \tag{3.2.15}$$

where

$$\boldsymbol{M}_{\mathrm{O}} = \sum_{i=1}^{n} \boldsymbol{r}_i \wedge \boldsymbol{F}_i^{\text{ex}} \tag{3.2.16}$$

represents the moment of the external forces about the fixed origin.

Relation (3.2.16) represents another important deduced hypothesis. Expressed in words it states that the moment of the external forces about the origin of an inertial frame equals the rate of change of angular momentum about that origin. To the preceding two examples one could add many other deducible hypotheses (for example, that (3.2.15) holds in corresponding form relative to the centre of gravity of the system—*cf.* Synge and Griffith (1942)—that for a conservative system, for which the forces can be derived from a potential field, the sum of potential and kinetic energy is constant, and so on). But the two illustrations will have to suffice to show how deduction calculi (vector analysis, differential calculus, and so on) are used in classical mechanics in order to derive deduced hypotheses from the initial ones.

3. As a typical example of an internal constitutive force acting between the particles of a point particle system, we can take Newton's postulate that the gravitational force which the jth particle exerts on the ith is given (see appendix for notation) by

$$\boldsymbol{f}_{ij} = \gamma m_i m_j(\boldsymbol{r}_j - \boldsymbol{r}_i)/|\boldsymbol{r}_j - \boldsymbol{r}_i|^3 \tag{3.2.17}$$

where γ is a universal gravitational constant.

4. If we consider two gravitating particles, forming a 'two-particle' system, which in *idealisation* is assumed to exist in isolation (that is, we neglect any external forces acting on the system), then equations (3.2.7) which, with the help of (3.2.17), now take the forms (since $\boldsymbol{F}_i^{ex} = 0$ for $i = 1, 2$)

$$m_1\, d^2\boldsymbol{r}_1/dt^2 = \gamma m_1 m_2(\boldsymbol{r}_2 - \boldsymbol{r}_1)/|\boldsymbol{r}_2 - \boldsymbol{r}_1|^3 \tag{3.2.18a}$$

and

$$m_2\, d^2\boldsymbol{r}_2/dt^2 = \gamma m_1 m_2(\boldsymbol{r}_1 - \boldsymbol{r}_2)/|\boldsymbol{r}_2 - \boldsymbol{r}_1|^3 \tag{3.2.18b}$$

can be solved by well-known methods (*cf.* Synge and Griffith, 1942; Synge, 1960). By considering (in idealisation) the sun and the planets each as point 'particles' (since their radii are very small compared to their distances apart), we can then derive from (3.2.18a, b) Kepler's three laws of planetary motion for *one planet* (*cf.* Joos, 1947; Synge and Griffith, 1942), which provide three typical empirical generalisations.

The preceding examples illustrate (i) that important empirical generalisations can be deduced from the initial hypotheses of mechanics (for other examples see below), (ii) that these derivations involve the use of deductive calculi, and (iii) that they involve idealisations or approximations. Better approximations could be obtained when considering the influence of other planets and satellites on the planet considered, by use of Hamiltonian perturbation theory.

5. As a final example of classical mechanics we consider the movement of the bob of a simple pendulum. Here again one idealises by making simplifying assumptions. It is assumed that the bob is a point particle of mass m, suspended from a fixed point O by means of an inelastic string whose mass can be neglected compared to that of the bob. It is also assumed that the initial conditions are such that the pendulum will oscillate in a vertical plane. It can then be deduced (for example, from (3.2.15) on taking moments about O)

that the equation of motion is

$$a\, \mathrm{d}^2\theta/\mathrm{d}t^2 = -g \sin \theta \qquad (3.2.19)$$

where a is the length of the string, g the earth's gravitational constant (in the locality considered) and θ the inclination of the string to the vertical at time t. Although (3.2.19) can be formally solved (in terms of elliptic functions), we obtain in the mathematical *approximation* of small oscillations (for which $\sin \theta$ can be approximately replaced by θ) that

$$a\, \mathrm{d}^2\theta/\mathrm{d}t^2 = -g\theta \qquad (3.2.20)$$

This differential equation has a periodic solution, with period $T = 2\pi(a/g)^{\frac{1}{2}}$.

The preceding example shows that by making *approximations* in the *deductive calculations* (in addition to idealisations) we can obtain a simple empirical generalisation, namely the periodic time as a function of the length of the string and of the local geographical value of g. When people make mathematical approximations, and treat certain terms as being small compared to others, there appears in many, if not all, cases an arbitrariness concerning what is to be regarded as 'small', in the sense that it can be discarded. The reason is that there exist no absolute criteria for deciding when one quantity is small compared to another one of the same physical 'dimensions' (the word 'dimension' is here used in the strictly technical sense of physical dimension theory, where for instance length, mass and time are regarded as each having different, mutually irreducible, 'dimensions'; *cf.* Bunge, 1967b, pp. 37–8). In practice scientists often agree that certain terms in equations can be considered as being small compared to others. Yet there is always room for improvement, and people are often simply forced to neglect certain terms in equations or to make certain idealisations (such as treating large bodies as if they were points, or introducing simple boundaries (spherical, say) of domains) in order to be able to solve differential equations (or algebraic equations), whose solutions may then approximate to particular, already known, empirical generalisations. (For instance, the 'Kirkwood approximation' used by Born and Green (1949) in the kinetic theory of liquids provides a typical example of an idealisation which results in a large reduction of the number of variables. Without making such idealisations—which amount to *ad hoc* hypotheses—many or most systems could never be treated in terms of existing powerful physical theories.)

It should be noticed that statistics enters into *classical* mechanics only at the stage where empirical generalisations are compared with experiments. For instance, Kepler's laws or the formula for the pendulum represent 'empirical generalisations', and statistical least-square fitting of data can ascertain how well these laws are approximated to by observed mean values of variables. In other theories statistics may be introduced in the form of probabilities into the initial hypotheses (as in genetics). In particular the introduction of probabilities into some of the initial hypotheses of quantum mechanics has been the subject of several partly radically different approaches and interpretations (see Born and Jordan, 1930; Moyal, 1949a, b; Fine, 1968; Margenau and Cohen, 1967; Born, 1949, 1961).

B. The Jacob–Monod theory of genetic regulation

Much of this theory is at present qualitative, and provides an excellent example of the manner in which molecular biologists can make verbal deductions (using technical language and concepts) from given initial hypotheses. A detailed survey of the essentials of this theory was given by Jacob and Monod (1963) and brief surveys are given by Wassermann (1972) and De Busk (1968, chapter 5) among others.

It is not possible to give here a detailed or formal account of the theory, as it would demand too much technical knowledge in the field of bacterial molecular biology. Suffice it to state that one can represent the theory in terms of initial hypotheses, which specify (a) machinery whereby DNA (deoxyribonucleic acid), which, by hypothesis, forms the genetic material of a bacterial gene, makes RNA (ribonucleic acid); and (b) machinery whereby particular types of RNA molecules, namely 'messenger RNAs' (mRNAs) cause the production of specific polypeptide chains, one or more of which form principal constituents of any particular protein. The Jacob–Monod theory is primarily concerned with mechanisms which allow the 'switching on' or 'switching off' of the machinery of type (a). The theory hypothesises the existence and mode of action of various types of regulator molecules: (i) inducers which can 'switch on', and (ii) repressors which switch off mRNA synthesis. From the initial hypotheses of the theory, one can make numerous deductions, leading to empirical generalisations which have been confirmed. Many of these deductions are of a negative kind, suggesting for example that mutations of the various postulated genetic control elements of the theory should affect the control of mRNA synthesis (for instance they suggest that mutation of a regular gene which caters for repressor molecules should affect mRNA synthesis).

The theory postulates properties of a number of hypothetical structural elements which, like forces in classical physics (or electron spin in quantum mechanics), cannot be directly observed, but are inferred 'unobservables', which are useful variables of hypothetical systems. For instance, the hypothesis that a repressor molecule acts on a postulated 'operator' of a gene introduces an unobservable 'operator' element whose *properties* can only be indirectly inferred by introducing operator mutations and 'genetic mappings'. Again, there is the hypothesis that an RNA from which mRNA is either processed or which already is an mRNA starts being synthesised at a specific region of a bacterial gene, which is referred to as the 'promoter'. The promoter forms (by hypothesis) the locus of initial action of a DNA-dependent RNA polymerase (an enzyme which step by step adds constituents to the growing mRNA precursor chain), and a promoter is another hypothesised structure of the theory. However, what matters is not the postulated structures but their *properties*. By means of artificially produced mutations one can test certain deducible consequences of these postulated properties (*cf.* Jacob and Monod, 1963).

3.3 General features of theories

Whenever a set of initial hypotheses—see (3.2.1)—has been established, theorists attempt to make as many deductions from them as possible. All deductions from initial hypotheses are 'deduced hypotheses', some of which may be empirical generalisations. For example, (3.2.10) and (3.2.15) are deduced hypotheses derived from the initial hypotheses of classical mechanics. Deduced hypotheses need not necessarily be empirical generalisations, and many are not. The deduced hypotheses of a theory may be ordered to form a set denoted by

$$[h_i; \text{for } i = 1, 2 \ldots] \tag{3.3.1}$$

Although at any one time in the history of a science set (3.3.1) has a definite number of members, the actual number may change, as time progresses, with the discovery of new deductions.

In works on the philosophy of science, it is usually stated that the h_is are derived from the H_is and from each other by means of deductive calculi (such as formal logic, vector algebra or matrix algebra, as illustrated by some preceding examples). However, in some cases (as in molecular biology and other sciences), the deductive 'calculus' used in certain types of deductions may be a purely verbal argument using the technical terms of the theory. In fact, in molecular biology sophisticated 'ordinary reasoning' is applied more often than not (*cf.* the paper by Crick *et al.* (1961) on the genetic code). For any theory the intricacy of the calculi and logical chains used depends on the state of development of the science concerned. But even in the most advanced theories (such as quantum field theories; *cf.* Schweber, 1964), verbal interpretations and arguments form important companions of deductive calculations. It must also be remembered that some h_is may be derived from other h_is or from H_is by procedures which involve mathematical *approximations* (or additional simplifying *ad hoc* hypotheses).

If a theory is to be empirically anchored some of its h_is must be empirical generalisations. The larger the number of its h_is which are confirmed empirical generalisations, the more firmly will a theory be empirically anchored. Many philosophers of science have stated that a reconstructed theory postulates *semantic rules* relating some, but not all, terms which occur in the H_is and h_is to the 'data language' which refers to publicly observable objects, situations and events. For instance, in classical mechanics some gravitational masses may, *via* semantic rules, be related to measurements which permit comparison of the gravitational mass of a given system with a recognised standard (as occurs, for example, in less sophisticated procedures relying on simple calibrated weighing machines). Although some theorists and philosophers of science (the so-called 'instrumentalists') like to regard theories only as *formal* instruments of deduction (see p. 77), whose variables could remain semantically undefined until one reaches the stage of deduction of empirical generalisations, I believe that this way of looking at theories is unprofitable. The fact that theorists often use idealisations in formulating initial hypotheses,[2] or rely on approximations (for example, by discarding terms) when using deductive calculi, makes semantic identifications of as many theoretical terms as possible either very desirable or indispensable at all stages.

A semantic interpretation of theoretical expressions often provides a valuable *heuristic guide* in deciding which terms in a (mathematical) theoretical expression can be dispensed with. Without having a pretty good idea of the *meaning* of terms they are handling, theoretical physicists and chemists would often not be able to make reasonable approximations in their theories. Theoretical physicists often use approximation procedures based on estimates of orders of magnitude. A theoretician may argue that a certain quantity can, for good reasons, be expected to be of negligible magnitude by comparison with other quantities (for example, because of analogies with other systems for which similar calculations were made and compared with experiments). To obtain such estimates one must know the semantic interpretations of theoretical variables or expressions which are being estimated.

Similarly, when molecular biologists theorise about genetic regulation, it is important for them to know the meaning of the terms involved. For instance, if we wish to estimate the number of protein molecules which could maximally be associated with a certain stretch of DNA, it is important to identify the protein types concerned (for example, to assume that they are histones), since proteins can vary considerably in molecular dimensions. Hence, in all kinds of theoretical estimates constant semantic identification of terms is highly desirable. If we want to estimate the packing density of cells in a given tissue, we must have a good idea of the types and sizes of the cells concerned—that is, theoretical packing densities may depend sensitively on the semantic identifications made.

These and many similar examples illustrate and suggest why in practice most scientists do not regard theories as mere instruments of calculations. Semantic identification does not allocate reality status to theoretical entities.

[2] For example, some postulated formulations of classical statistical mechanics of gases assume that gas molecules behave like 'hard spheres' and are perfectly elastic.

For instance, when an electronmicrograph indicates that a certain intracellular strand (say a presumptive mRNA) 'carries' a string of n blobs, then the semantic identification of these blobs with hypothesised ribosomes (which are particles of *ca* 100 Å diameter; for composition see Wassermann, 1972, chapter 5) does not mean that one has 'seen' ribosomes. What one has seen are blobs in an electronmicrograph (*cf.* p. 52) and nothing else; the semantic identification is and remains an interpretation. (Compare this with the following passage from Oscar Wilde's *Salome*. '*Herod*: The moon has a strange look tonight. Has she not a strange look? She is like a mad woman, who is seeking everywhere for lovers. She reels through the clouds like a drunken woman . . . . *Herodias*: No, the moon is like the moon, that is all.')

3.4 The nature of explanations

The scheme of figure 3.1 illustrates the typical *network structure* of initial and deduced hypotheses of a theory. In this scheme some (but not all) of the

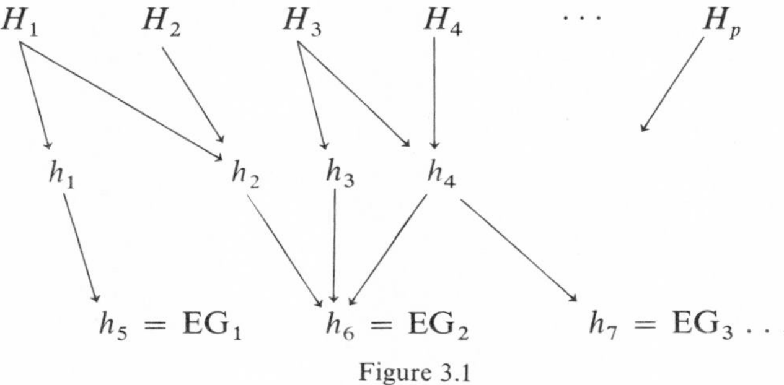

Figure 3.1

deduced hypotheses, namely (h_5, h_6, h_7 . . .), are empirical generalisations (denoted by EG for short). It is seen that any empirical generalisation which is deducible from the theory may, *via* intermediate h_is, be linked to one or more initial hypotheses. Thus, $\mathrm{EG}_2 = h_6$ is linked *via* h_2 with H_1 and H_2, *via* h_3 and h_4 with H_3 and *via* h_4 with H_4. Conversely, any one of the initial hypotheses may be linked *via* intermediate deduced hypotheses with a plurality of deduced EGs. Consequently if, say, $\mathrm{EG}_{j_1}, \mathrm{EG}_{j_2} \ldots \mathrm{EG}_{j_t}$ are EGs which all depend on deductions from the initial hypothesis H_m when these deductions are appropriately combined with deductions from other initial hypotheses—say H_k; for $k = k_1, k_2 \ldots k_r$, where $k_i \neq m$ for $i = 1, 2 \ldots r$—then any empirical evidence which confirms, say, EG_{j_i} (for $1 \leqslant i \leqslant t$) could indirectly help to strengthen the plausibility of H_m from which EG_{j_i} is deduced *via* intermediate deduction hypotheses.

Alas, the matter is not quite as simple as this. If the intermediate deductions are *rigorous*—that is, if they avoid discarding terms in mathematical equations or in mathematical expansions, and involve no computations with potential accumulative errors (say in repeated numerical integrations)—then the logic of deductions often discussed by philosophers of science is perhaps applicable. However, if mathematical approximations (such as the discarding of terms) and accumulative errors occur, then the deduction of EG_{j_i} could be partially fortuitous. It could resemble the sad state of affairs, well known to mathematics examiners, that candidates may obtain correct results from totally faulty premises by making compensating deductive mistakes.

For these reasons Popper's falsification arguments also require great care in actual applications. If we derive one or more EGs from several initial hypotheses of a theory and these EGs are at variance with facts, this does not necessarily mean that either (a) all, or (b) even any of the initial hypotheses used are at fault. It could signify either (i) that one or two of the initial hypotheses are at fault, or (ii) that none of the initial hypotheses is inappropriate, but that in applying deductive calculi approximations were made and computational errors accumulated which led to faulty EGs. (Approximations in quantum mechanics are widely used: for example perturbation theory (Kato, 1966; Wassermann, 1946, 1947, 1949) or variational approximations (*cf.* Kemble, 1937, p. 579; Harrison, 1970, pp. 499ff; Dicke and Wittke, 1963, chapter 14)). Use of better calculations might then lead to the deduction of empirically adequate EGs.

In subjects like quantum chemistry, where sizeable molecules are sometimes treated quantum-mechanically by various mathematical approximation procedures (*cf.* Coulson, 1961; McWeeney and Sutcliffe, 1969), any disagreement of calculated binding energies of molecules, or of calculated bond angles and bond lengths and the like, need not be inherent in the general quantum-mechanical theory which is being used, but could be due to idealisations (simplifications of general equations of the theory), mathematical approximations and accumulations of computational errors. On the other hand, if a given method of approximation gives agreement for a particular molecule, this is no guarantee that some of the agreement of calculated values of various parameters could not be fortuitous, unless similar types of calculation lead to equally good agreement with experiments for many different types of molecules.

In general a theory is said to *explain* a set of empirical generalisations when these can be deduced from its initial hypotheses. The plausibility of such explanations depends then on the forcefulness (for example, degree of absence of approximations and inherent errors) of the deductive procedures which may have been used. It is also seen that a theory may often serve *indirectly* to link up a set of empirical generalisations which are derivable from that theory and which are in agreement with data to an acceptable degree of approximation. In this manner, as the scheme at the beginning of this section shows in graphic form, evidence for each empirical generalisation which is derivable from the theory can *potentially* be regarded as providing (*via* the initial hypotheses) indirect evidence for one or more other empirical generalisations of the theory. Inclusion of the word 'potentially' in the preceding sentence severely differentiates the present interpretation of the function of hypothetico-deductive theories from the related interpretation by Braithwaite (1953), who treated deductions as if they were *exact* logical procedures. However, as theories make often many simplifications in their deductive calculi and in the actual use of their fundamental hypotheses, one often cannot consider their deductions as being *logically* adequate for assessing their premises. Hence explanations may involve a good deal of personal judgment (concerning the validity of deductions and initial hypotheses), even if they can be corroborated by approximate agreement with numerous different EGs (compare the many partial successes of the Bohr–Wilson–Sommerfeld rules of the older quantum theory).

In one respect Braithwaite (1953, p. 18) is partially correct:

> no amount of empirical evidence suffices to prove any one of the hypotheses in the system, yet any piece of empirical evidence for any part of the system helps towards establishing the whole of the system. (See also Hull, 1943, p. 12.)

As no finite amount of evidence suffices to establish an EG, it is clear that no finite amount of evidence suffices to prove any hypotheses of a theory, since the latter can only be empirically supported *via* EGs. The second part of Braithwaite's assertion is not necessarily valid, as he has disregarded the approximate nature of many deduction procedures used in practice. In many qualitative theories certain hypotheses are only very weakly empirically anchored *via* an indirect chain of arguments. For instance, the hypothesis that protein synthesis depends on the movement of ribosomes relative to mRNA strands, though *consistent* with many facts, has not been demonstrated by direct evidence which bears on the movement *per se*. Nevertheless the hypothesis, combined with the postulated action of mRNA and of aminoacyl-transfer-RNAs (see Wassermann, 1972) explains many confirmable aspects of polypeptide chain synthesis (including properties of polar mutations). The hypothesis is therefore valuable but not yet as 'strong' as the hypotheses of quantum mechanics which implicate (equally unobservable) movements of electrons (and which, *via* spectroscopy and chemistry are very firmly empirically anchored through a sequence of intermediate deductions).

The preceding discussions were not intended to disparage hypothetico-deductive theories (as was attempted by Harré, 1970, pp. 8ff). My conclusions undermine neither their explanatory nor their predictive value (see also § 3.13), otherwise I would be trying to saw off the branch on which I am sitting. Hypothetico-deductive theories have achieved remarkable unifications and predictions in innumerable instances. A few examples should suffice to remind the reader of this.

(i) Maxwell's electromagnetic theory (*cf.* Stratton, 1941, for a systematic account) has explained a wide range of phenomena concerning conduction of electricity, and related magnetic properties, and its predictions form the background of much of current electrical engineering practice. Apart from this it provides the deductive basis for the theory of optical phenomena (*cf.* Born and Wolf, 1959, and recent volumes in *Advances in Optics*, edited by E. Wolf).

(ii) Classical mechanics has scored by the explanation of (a) Kepler's law, (b) the behaviour of Foucault's pendulum, and (c) a host of other important phenomena, apart from having a great variety of engineering applications, and similar remarks apply to classical continuum mechanics (linearised elasticity theory, for example).

(iii) Quantum mechanical explanations of (a) atomic structure and the properties of the periodic system (*cf.* van der Waerden, 1932), (b) the nature of chemical bonds (*cf.* Coulson, 1961; Heitler, 1945), (c) electric conductivity in metals, including superconductivity, and (d) interpretations of a wealth of spectroscopic data (*cf.* Condon and Shortley, 1935), provide some of the best known examples of the unificatory powers of hypothetico-deductive theories.

(iv) The Gell-Mann phenomenological theory of elementary particles and current algebras (*cf.* Renner, 1968) illustrates another successful integrative theory which had important predictive consequences.

(v) X-ray diffraction theory has formed a rich source of information concerning hypothesised structures of a great variety of simple molecules, and more recently even of macromolecules, including DNA and proteins (*cf.* Wassermann, 1972, for a brief survey of DNA and protein structure, and Blake *et al.*, 1972, for a (typical) recent structure determination of the protein phosphoglycerate kinase of horse muscle; for the structure of numerous other proteins, see *Cold Spring Harbor Sympos.*, vol. 36).

(vi) Molecular biology is now fast catching up in suggesting explanatory unifications in biology and psychobiology (e.g. Wassermann, 1972, 1974).

My arguments concerning hypothetico-deductive theories only suggest that in many, or most, cases mathematical logic cannot be used (a) as a method of formalising (or simulating) deduction procedures in empirical sciences and, hence, (b) in attempts to *justify* any relations of truth or falsehood between initial hypotheses and derived empirical generalisations within theories. This applies particularly when either human judgments or complex mathematical approximation procedures (such as perturbation theory), which do not rely on successive approximations, are used. This brings me back to Harré's (1970) views referred to above, with which I do not agree.

His views fail to separate (a) the practice of deductive scientific theorising from (b) the attempts by philosophers of science to 'justify' these practices by means of formal logic. I have argued repeatedly that, since approximate scientific deductions and idealisations involve valuations and decisions, no logical 'justification' of deduction procedures can be envisaged. Even if we accept a logic of inexact classes for empirical generalizations (Körner, 1966; Cleave, 1970), it is far from clear that this logic is or can be related to the mathematical approximation procedures of mathematical calculi used in deducing EGs from initial hypotheses of a theory. Körner is dealing with *inductive* properties of EGs and not with the manner in which EGs are deduced within theories.

Harré (1970, p. 96) believes that

> It is obvious that, using classical logic, the truth of consequences cannot be transferred to their antecedents, because on the deductive model, infinitely many theories could be constructed to imply the same consequences, only one of which might be true. The *reductio ad absurdum* of deductivism comes from seeing that, in accordance with the canons of received logic, the only relation theory can have with fact is that a theory or hypothesis is falsified by the discovery of states of affairs the description of which contradicts conclusions from the theory.

The circumstance that empirical generalisations cannot be proved to be true by (a finite set of) data, makes it irrelevant that 'the truth of consequences cannot be transferred to their antecedents', since nobody could claim that any scientific hypothesis (not even an empirical generalisation!) can on empirical evidence be shown to be true (in the sense of formal logic). (See also my remarks on 'falsifiability' on pp. 32–34.) It has long been recognised that people might conceivably find totally different hypothetico-deductive hypotheses which cover the same range of empirical generalisations as some existing theory. But this lack of demonstrable uniqueness of initial hypotheses is basically no different from the lack of uniqueness of least square fits. Moreover, in cases where different theories nearly led to similar EGs (compare, for example, modern quantum mechanics with its predecessor based on the Hamilton–Jacobi theory reinforced by the Bohr–Wilson–Sommerfeld rules: *cf.* Born, 1927, for a lucid exposition of the older theory), the later theory proved much superior in accuracy and explanatory range (see p. 77). Every working theoretician knows how hard it is to discover any theory which spans a wide range of phenomena in terms of relatively few initial hypotheses (for example classical mechanics, electromagnetic theory and quantum mechanics).

It is possible that future theories may be found which comprise existing ones as approximate cases (much as contemporary quantum mechanics can explain some formulae of the Bohr–Wilson–Sommerfeld theory as applications of the Wentzel–Kramers–Brillouin (WKB) approximation: *cf.* Dicke and Wittke, 1963, p. 245). But there can be no guarantee that alternative theories might not be discovered which cover the range of phenomena of an existing theory equally well and no others; that is, we cannot give uniqueness proofs for theories, nor prove their necessity (see Toulmin, 1971). At best we can demonstrate the sufficiency of a hypothetico-deductive theory for explaining or predicting a wide range of phenomena. Scientists are satisfied with the sufficiency of theories, but some philosophers of science seem to believe that necessity and uniqueness proofs are essential and that theories must be logically 'justified'. It is true that for an established theory it may be important to show that any possible *exact* deduction from initial hypotheses is unique. *However, this is not the same as demanding a proof that the theory itself is unique.*

A theorist must also give reasons for rejecting known alternatives in preference to his own hypotheses. Resort to empirical evidence may help to eliminate as many potentially equally plausible alternative explanations (if these are known or would appear reasonable to those working in that branch of science). Also, new crucial experiments may have to be proposed. Long-lasting ambiguities of interpretation often arise (*cf.* Anon, *Nature*, 1972, **235,** 360–1 for a good recent example from molecular biology). In general, the wider the range of confirmable deducible EGs of a theory (which are obtained without additional *ad hoc* hypotheses), the more credible the theory becomes (by comparison with competitors, or on its own).

Harré's approach also disregards that linkage of deduced EGs *via* a network of hypotheses of a theory is often not equivalent to a formal logical linkage. These linkages are not concerned with interrelating truths or falsities of propositions. In fact I have stressed the dangers which confront Popper's point of view if one wishes to use these linkages for falsifying an initial hypothesis of a theory (see p. 92). Scientists may claim that from specified initial hypotheses they can deduce particular EGs, but they cannot claim demonstrability of the logical truths of either initial hypotheses or of EGs, and hence do not have to establish what they do not claim. We say: if a deduced EG clashes significantly with experimental data, then we must think again. Maybe a mathematical approximation misfired, or one or more of the initial hypotheses have to be replaced by better ones. Thus, scientific theory construction is *a trial-and-error business* and not a formal logical algorithmic procedure. There exist no rules for discovering appropriate initial hypotheses and deduced hypotheses, any more than there exist rules for discovering an invention or for any other act of creativity. Acts of creativity may have to conform to certain rules (as when we use mathematics), *but* these rules do not guide us towards a best possible choice of hypotheses.

Even if philosophers of science could describe approximation *procedures* in terms of appropriate formal logical tools and allocate measures (Carnap-type 'probabilities') in order to assess the degree of confirmation of initial hypotheses (see § 3.11) of a hypothetico-deductive theory, they would not get any further. They would still require arbitrary social decisions (as in the case of statistical 'significance tests') as to which probability values are to be taken as adequate for accepting or rejecting an initial hypothesis. People could argue that uniform criteria of acceptance must not be used, particularly in the case of less plausible types of hypotheses.

We realise therefore that science is a social enterprise (*cf.* Ravetz's (1971) long discussions of certain relevant aspects) necessitating ultimate agreements among scientists. But this presents the new danger that existing orthodoxy makes a collective misjudgment. For this reason (particularly if a theory introduces drastic new hypotheses such as Planck's quantum hypothesis, Einstein's theory of relativity, Mendel's hypotheses) it may take a long time for a new theory to be accepted. Many arguments may take place behind the scenes or even in public (for example the Darwin–Wilberforce controversy concerning evolution) before new hypotheses are taken seriously, particularly if they threaten to overthrow well-entrenched hypotheses which have, in the minds of some scientists, mistakenly acquired the status of absolute truths.[3] This, alas, is the source of scientific dogmatism. The prestige of a recognised authority (such as Dalton's followers *versus* Avogadro) may sway scientific social opinion for a long time. However, sooner or later scientists almost invariably recognise the significance of important major theories. Mendel's theory of inheritance was dispatched into the waste paper baskets of many of his better known contemporaries, being left unread because he had committed the sin of being neither well known nor attached to a great centre of learning. For instance, Nägeli, on receiving a copy of Mendel's manuscript, sent him a patronising letter. But ultimately Mendel's work was recognised and greatly outshone anything Nägeli had achieved.

When scrutinising new hypothetico-deductive systems, scientists may require great skill in making plausibility assessments, and different scientists (depending on their individual skill and knowledge) may greatly vary in their opinions regarding such work. Alas, even personal motives may enter (see, for example, Lenard's assessment of Einstein's theory of relativity in the former's book *Deutsche Physik*, where vicious antisemitism was the primary motive).

3.5 'Unobservable variables' and extensible theories

Although all initial hypotheses of a theory could, in principle, be lumped together into a single hypothesis, their splitting into a set of *separate* initial hypotheses of type (3.2.1) often permits an *extension* of a theory. For instance,

[3] For a particularly illuminating example of the use of *a priori* arguments by scientists against new discoveries, the reader is referred to Hempel (1966a, p. 48) for an account of the arguments given by the astronomer Sizi against Galileo's discovery of the satellites of Jupiter.

by postulating as some initial hypotheses of molecular biology that (a) ribosomes move along RNA strands of certain kinds, and (b) they participate in protein synthesis *via* additionally assumed mechanisms (*cf.* Wassermann, 1972, chapter 6), people could *extend* these hypotheses and these mechanisms to (i) synthetic types of RNAs (polyribonucleotides) operating in cell-free systems, (ii) to viral RNA, and (iii) to all kinds of native 'cellular' mRNAs. Again, the initial hypothesis (3.2.2) of Newtonian mechanics can be combined with a wide variety of hypotheses which provide definitions of constitutive forces for various systems. By eliminating the force $\boldsymbol{F}$ between (3.2.2) and the equation defining $\boldsymbol{F}$ in terms of the constitutive parameters of the system which acts on a particle, one obtains a differential equation for the motion. In this way classical mechanics can be applied to many different systems acting on a point particle. Suitable illustrations can be found in textbooks on classical mechanics (*cf.* Synge and Griffith, 1942).

Another example of an extensible theory is Maxwell's classical electromagnetic field theory. The theory imposes certain boundary conditions at surfaces bounding the system concerned. For instance, when applying the theory to waveguides with different boundaries, one can obtain a wide variety of derived EGs, each appropriate to a particular boundary, and the theory can be *extended* (and tested) for a virtually unlimited variety of boundaries. Again, when applying Maxwell's theory to optical imaging, various combinations of lenses of different shapes can be considered. Quantum mechanics has also been *extended* to embrace a wide variety of phenomena ranging from explanations of the properties of superfluids (liquid helium and superconductivity) at one extreme to applications in quantum chemistry and nuclear physics (and the like) at the other.

In order to *combine* individual mathematically formulated initial hypotheses, it is essential to introduce unobservable variables. For instance, in classical mechanics forces are such (eliminable) unobservables (*cf.* P. Frank's (1947) extensive discussion of this and related points). Some scientists would argue that forces are observable since according to (3.2.2) they can be defined (for a point particle) in terms of acceleration ($\mathrm{d}^2\boldsymbol{r}/\mathrm{d}t^2$) and mass m. But such assertions are doubtful in many cases. For example, in the case of planetary motion we cannot directly determine the mass m of the sun by comparing it with a standard mass. It is only by combining (3.2.2) with Newton's gravitation hypothesis (3.2.17) that, on eliminating the forces, we can obtain an equation of motion whose solution allows us from observation of other variables (which appear in the solution) to deduce the approximate value of the mass of the sun. Again, applying similar consideration to the earth and the moon, we could obtain the approximate mass of the earth.

In quantum mechanics, Schrödinger functions are unobservables, and the same applies to the field variables of electromagnetic theory. However, by *averaging* over these unobservables one can derive expressions in which the unobservables no longer appear (see p. 58 above). In these cases unobservables are therefore 'eliminated' through statistical procedures.

In *behavioural science* unobservable variables are usually referred to as '*intervening variables*' (Tolman, 1938). Some psychologists believe mistakenly that the introduction of separate H_is and of *unobservables* into behavioural science is only a matter of convenience. They argue that this provides a 'useful way of breaking down into more manageable form the original' empirical generalisations (E. C. Tolman, 1938, p. 9). However, this interpretation of theories misses the *essential necessity* of introducing unobservables (or intervening variables) if (i) *several* EGs are to be linked *via* initial hypotheses, and (ii) the theory is to be *extensible* (see also Hempel, 1958). The *number* of initial hypotheses of an extensible theory need not necessarily be less than the number of EGs deducible from the theory (*pace* Koch, 1954, pp. 25–6). For instance, the theory of liquids by Born and Green (1949) established a formidable array of initial hypotheses (H_is) but encountered at the time of its formulation great difficulties in making adequate deductions from these. As time progresses, the ratio of EGs deduced from a theory (originally poor in deduced and confirmed EGs) to the number of initial hypotheses may increase. Moreover, in the light of experience some, but not necessarily all, of the initial hypotheses of a theory may have to be replaced by others which allow the deduction of EGs which are in better agreement with experimental data.

A theory often uses, apart from its initial hypotheses, deduction calculi (such as formal logic, vector analysis, differential and integral calculus or matrix algebra), the consistency of whose axioms may not be provable within that calculus (see Gödel, 1930a, b, and pp. 23–24 above). Any inconsistencies among derived EGs, or disagreement of EGs with observations, could in principle be due (i) to inconsistencies among the axioms of the calculi used, (ii) to an inconsistent choice of initial hypotheses of the theory, (iii) to both (i) and (ii), or (iv) to faulty approximations, accumulations of error, and so on. A single inappropriate initial hypothesis of a theory could lead to the deduction of a sizeable set of EGs which are at variance with observations. (Again, the approximate use of a particular calculus may introduce serious errors. For instance, when employing slowly converging mathematical series and breaking these off too soon, we could get inappropriate results without the basic theory being at fault.) It must also be remembered that empirical agreement with any particular EG derived from the initial hypotheses of a theory can (as a rule) only be considered as evidence in favour of the combined system of (i) initial hypotheses, (ii) approximations and computational methods, and (iii) the axioms of the calculi chosen (prior to approximation).

In some 'semi-empirical theories' the initial hypotheses may contain unknown parameters which are fitted to empirical data only at the stage of

EGs. For instance, as mentioned, the mass of the sun enters equation (3.2.2) as an unknown mass parameter which remains undetermined until at the stage of deduction of Kepler's laws one can, by extrapolation from empirical data, approximate to the mass value. In some branches of nuclear physics people introduced potential functions which depended on undetermined parameters which were ultimately fitted to empirical data. In quantum mechanical calculations this is not uncommon practice (*cf.* Bunge, 1967b, p. 76, for arguments relating to the inherent dangers of this procedure).

If several different theories, based on different sets of initial hypotheses, should make it possible to derive the same set of EGs, and no additional EGs which differ and which can be submitted to a 'crucial test', then the principle of economy of initial hypotheses could be invoked so as to favour that theory which has the least number of initial hypotheses (but see p. 72, note 9). This principle is neither logically necessary nor always as obvious to apply in practice as some philosophers of science believe (see also Köhler's (1938) discussion; Berenda's (1950) general appraisal of conventions, and Woodger's (1956, p. 60) comments on 'bolstered up' theories). In fact, sometimes a hypothesis is a single statement—(3.2.2), for example; in other cases it consists of a 'battery' of statements which may comprise a set of implicitly or explicitly stated relations.

The preceding brief discussion of some of the formalistic aspects of theory construction may be wound up with Freeman's (1948, p. 3) admirable assessment of the enterprise:

> Neither psychology, nor any other science, can escape the need for theory construction. A theory exists to bridge gaps in knowledge. It serves to organize scattered phenomena into a consistent whole. It proposes questions of alleged relationship and opens these to experimental attack. In short, theory develops rather than impedes scientific progress.

3.6 Emergentism *versus* reductionism

A. Hierarchies of theories and metatheories

As the initial hypotheses (3.2.1) of the theory T (say) are *postulated within* that theory, they remain unexplained as far as theory T is concerned. However, scientists often construct metatheories such that metatheory $M(T)$ permits them to deduce from its initial hypotheses the initial hypotheses of (i) theory T, and (ii) *several other* theories T', T'' . . . . Sometimes several theories based on independent initial hypotheses have to be combined in order to deduce a plurality of other theories. For instance, I mentioned earlier that statistical mechanics, when combined with quantum mechanics and electromagnetic theory, makes it possible to deduce Ohm's law, the equations of states of idealised liquids, and many aspects of superconductivity (and so on). In general there could be an ascending hierarchy of metatheories, such that the initial hypotheses of several nth order theories can be deduced from the initial hypotheses of one $(n + 1)$th order theory. For example, the approximate deduction of the initial hypotheses of thermodynamics from those of quantum statistical mechanics (*cf.* R. C. Tolman, 1938; Born and Green, 1949) provides an illustration of an (attempted) deduction of a hierarchically lower order (phenomenal theory) from a higher order (molecular theory). The deduction of the phenomenological Ginsburg–Landau theory of superconductivity from the quantum-mechanical Bardeen–Cooper–Schrieffer theory (see Harrison, 1970, p. 527) gives another good illustration of the deduction of a lower order from a higher order theory. Again the approximate deductions of many aspects of the theory of superconductivity from the hypotheses of quantum mechanics (*cf.* Harrison, 1970, for a review) show how one theory can be deduced from another one. When this is the case people often speak of the *reduction* of a theory. Thus, superconductivity theory can be said to have been reduced to quantum mechanics. Likewise theoretical molecular physics forms a reductive metatheory for theoretical molecular biology.

A metatheory is redundant if it permits only deductions of the initial hypotheses of a *single* lower order theory *T*, without permitting deductions of EGs or of initial hypotheses of other theories which are not directly deducible from *T*. If it were not for this redundancy, an infinite regress of metatheories could be demanded.

Vitalists, organicists, holists and others have often alleged the futility of mechanistic theories, because these could never give *complete* explanations. It was argued that mechanists are always forced to retreat step by step in a 'causal chain'. Yet this argument is based on a complete misunderstanding of the formal structure of mechanistic theories. These start with initial hypotheses, which need not be explained within the theories. Mechanistic theories simply do not aim to give *complete* explanations, as their initial hypotheses remain unexplained within the theory. Ryle (1949), in his discussions of the 'bogey of mechanism', apparently did not appreciate this inherent incompleteness of mechanistic theories. In fact, the types of explanations which rely on non-terminated chains of 'causes' (infinite regress) are not invoking properly formulated scientific theories, and many laymen fall into this trap. Some people will ask 'how did Smith get the measles?' The reason given is that Smith got infected by Brown. It is then asked 'how did Brown get the measles?' The reason given is that Brown got infected by some other individuals and so forth indefinitely (*cf.* Hempel, 1966a, p. 58). A scientific explanation, however, runs differently. The scientist would start with the *initial hypothesis* that Smith's body became invaded by a dosage of measles virus. He would then argue that this virus replicated according to certain postulated mechanisms of molecular biology (*cf.* Wassermann, 1972) and caused lysis of cell components. From whom Smith received the initial dose of virus is (usually) totally irrelevant for the pure scientist, who is interested in explaining how viruses replicate and what particular effects they have on tissues, and how they can be neutralised or eliminated. ('Latent viruses' may require geographical distribution studies.)[4]

B. Reductionism and emergentism

It has often been claimed that certain theories are ultimately irreducible, that is, that their initial hypotheses will never be deducible from a known or still-to-be-discovered metatheory. Such prophecies are based on the bold assumption that their proponents can predict once and for all the outcome of all future discoveries and theories. Driesch (1929), for instance, convinced himself that it would be *a priori* impossible to explain morphogenesis in terms of mechanisms only. (I have recently suggested a molecular gene control theory of morphogenesis (Wassermann, 1972, 1973, 1974) which provides explicit mechanistic explanations of many discoveries of experimental embryology which Driesch thought to be *a priori* inexplicable in terms of mechanistic principles. Irrespective of the modifications which this theory may require, it provides an 'existence theorem'—that is, it suggests in detail that molecular biological principles suffice for the explanation of morphogenesis. As the mechanisms of molecular biology are in principle reducible to mechanisms of organic chemistry, and hence of quantum chemistry, it is seen that quantum mechanics should suffice as a metatheory for explaining the principles of morphogenesis.)

Some philosophers of science have alleged that various theories involve 'emergent properties'. This amounts also to an assertion that the initial hypotheses of such theories cannot be deduced from those of any possible metatheory. An often quoted example is the irreducibility of classical electromagnetic theory in terms of Newtonian mechanics. However, there exist no valid reasons for suggesting that Newtonian mechanics must be the appropriate metatheory for Maxwell's theory. Hence it cannot be argued that Maxwell's theory represents 'emergent' features not present in Newton's theory. In fact, many earlier attempts to construct a 'unified field theory' were based on the notion that there exist only electromagnetic and gravitational interactions in nature, and that all field phenomena ought to be deducible from a common metatheory, which yields electromagnetic theory as well as gravitational theory. However, in the meantime many new elementary particles and new types of interactions have been discovered (*cf.* J. D. Jackson, 1958; Renner, 1968; Schweber, 1964) which could be associated with a substantial variety of different types of fields. The fields hitherto suggested share a general formalism (for instance, their field equations are derivable from Lagrange density functions, and the fields are

[4] See G. Dean, 1970, *Scientific American*, **223**, no. 1, 40.

(at least on principle) quantisable, so that their 'quanta' represent the elementary particles associated with the particular fields—*cf.* Schweber, 1964; Visconti, 1969).

It remains possible (as Max Born suspected long ago in his 'reciprocity theory' which proved inadequate) that one day we may find a general field theory which allows us to deduce the field Lagrange functions of all elementary particles which occur in nature from general principles. Such a theory could serve as a metatheory for all 'particle theories'. Whether people will succeed in discovering such a Lagrange function generator theory, or whether other (as yet unsuspected) metatheories will be constructed, remains to be seen. However, Maxwell's theory is likely to form a special case of one of these metatheories. Moreover, the particles postulated in Newtonian mechanics are inadequately specified even from the point of view of contemporary elementary particle theories. Hence we could hardly expect that Newtonian mechanics could be derivable from a future metatheory for elementary particles (while this may apply to Maxwell's theory which when quantised—*cf.* Schweber (1964)—yields photons, that is, one important class of elementary particles.)

3.7 Persisting validity of theories

It is sometimes believed that the discovery of a new theory puts an end to a similar existing theory. While true in some cases, it is not universally true. For instance, when classical mechanics was found to be inapplicable to the phenomena of atomic physics, this did not mean that engineers could no longer confidently use classical mechanics within the confines of its *claimed* validity. The claims of validity of a theory depend on the range of its empirical confirmation. Thus, non-relativistic quantum mechanics is sufficient for much or all of quantum chemistry, but inadequate when dealing with high energy particle physics. Again, classical electromagnetic theory is good enough for engineering applications, but not for processes which can only be explained in terms of the quantum theory of the electromagnetic field (*cf.* Heitler, 1954; Schweber, 1964; Visconti, 1969).

3.8 Hypothetical constructs and 'pictorial' representations

I have pointed out that in many cases the independent and dependent variables of several empirical generalisations (such as Kepler's laws) become related *via* unobservable variables (such as the 'forces' of classical mechanics) which appear in the network of hypotheses of a theory. In general let us suppose that there are k independent variables

$$[x_i;\ \text{for}\ i = 1, 2 \ldots k] \tag{3.8.1}$$

k' dependent variables

$$[y_i; \text{for } i = 1, 2 \ldots k'] \tag{3.8.2}$$

and s unobservable variables denoted by

$$[u_i; \text{for } i = 1, 2 \ldots s] \tag{3.8.3}$$

Suppose further that there exist t relations of the form

$$[f_j(x_1, x_2 \ldots x_k; y_1, y_2 \ldots y_k; u_1, u_2 \ldots u_s) = 0; \text{for } j = 1, 2 \ldots t] \tag{3.8.4}$$

which make it possible to eliminate all unobservables of set (3.8.3), thereby permitting us to establish theoretically derived empirical generalisations. In theoretical molar *psychology* the unobservables are often referred to as 'intervening variables' (since in some 'black box' models they intervene between the 'input'—that is, the independent stimulus variables—and the 'output'—the dependent behaviour variables; see p. 65). In molar psychology the nervous system is treated as a 'black box' and, although intervening variables are sometimes given neurophysiological-sounding names, they are in black box theories not related to any neurophysiological discoveries.

When discussing a hypothetical entity (such as an electron) or a hypothesised event (such as a state transition of an atom) it may be inconvenient to think all the time of the numerous postulated relations and unobservables which characterise the entity or event. Instead it is convenient to refer to the entity by a generic expression (such as 'electron' or 'photon') implying that a great many initial hypotheses of appropriate theories (and their deductions) could be stated if desired, and are known to experts. Generic language is also encountered among molar behaviourists who are treating organisms as black boxes. They claim not to be concerned with *observing* structures and processes which reside or take place respectively 'under the skin' of animals or man. They treat a brain as (being for *them*) as *unobservable* system, but are prepared to attribute hypothetical effects to brains and other (for them hypothetical) systems which reside under the skin.

By the molar behaviourist the brain is regarded as a hypothetical construct, that is, a generic entity about whose postulated behavioural implications much could be said. But, in accordance with their methodology molar theorists believe that statements about brains should be made by neurophysiologists and neuroanatomists (who consider a brain no more 'hypothetical' than their laboratory equipment, although they formulate detailed *hypotheses* about its components, such as neurons (Hodgkin, 1964), synapses (Eccles, 1964a) and so on). It is this short-sighted compartmentalisation which, I believe, has led to the failure in psychology to establish proper theories. If physicists had similarly refused to link elementary particle physics and the physics of atoms and atomic nuclei (and so on) with molar phenomena, theoretical physics would still be in a poor state. Generic scientific terms (such as 'electron', 'proton', 'brain' or 'synapse') usually refer to systems or events with many hypothesised properties. These terms serve therefore as shorthand for sets of hypotheses (and deductions). A generic system whose properties are *purely* hypothetical within some particular theory (for example, the 'brain' within a molar behaviour theory, or an 'electron' in atomic physics) may be called a *hypothetical construct*, provided (a) that the generic system is adequately specified by defining statements (for example, of type (3.8.4), or of a verbal type), and (b) that from the defining statements *empirical generalisations* can (directly or indirectly) be deduced. This definition of a hypothetical construct does not necessarily coincide with that proposed by others (e.g. McCorquodale and Meehl, 1948; Bunge, 1967c, part 1, pp. 93ff).

A few examples of hypothetical constructs may help to illustrate their usage. When physicists speak of an 'electron gas in a metal', they are referring implicitly to a system of equations (such as the equations of Bloch's theory of metallic conduction: *cf.* Wilson, 1958; Harrison, 1970) which describe the postulated initial and deduced hypotheses, including empirical generalisations (such as Ohm's law or the Wiedemann–Franz law) together with semantic interpretations of the variables which characterise the system. Likewise a 'semiconductor', if considered as a *theoretical* system (*cf.* Peierls, 1955, chapter 10; Harrison, 1970, pp. 270 ff), refers to a set of hypotheses and derived empirical generalisations formulated *via* a system of equations. Similarly, Dirac's (*cf.* 1958) equations of the spinning electron and certain other hypotheses and their deductions may be taken collectively, together with semantic interpretations, as defining the class of statements which characterise the hypothetical construct 'electron' within that particular theory.

Again, when a theorist speaks of a 'hydrogen bond' he may have in mind a general theory of such bonds including all their postulated or deduced properties (*cf.* Orgel, 1959), the words 'hydrogen bond' serving as an abbreviation for a set of statements. Generic expressions such as 'Mendelian inheritance' or 'Lorentz invariance' are used by workers who know that these refer to a set of postulates of genetics and to specific transformation properties in the special theory of relativity, respectively, together with numerous interpreting statements. Useful as such shorthand terms may be, it is important to remember at all times the initial hypotheses of theories on which they are based.

I mentioned earlier (p. 61) that the hypothesis of repeat editions of certain (not directly observable) systems forms an important aspect of science. Terms such as 'electron', 'atom', 'hydrogen molecule', 'gene' (in molecular biology), and many others, refer each to a postulated repeat hypothetical construct which is specified by a set of initial and derived hypotheses (including EGs), the precision of which may depend on the state of the science concerned. (Hypothetical constructs imply what Bunge (1967c, vol. 1, p. 395) called conceptual connectedness, that is, that a set of hypotheses refer to the same hypothesised system or class of systems.) While the system itself may be unobservable, some consequences of its postulated hypotheses must be amenable to observation. There may exist a hierarchy of hypothetical constructs (such as elementary particles, atoms or molecules) and hence a hierarchy of conceptual connections within a theory.

The definition of a hypothetical construct can change if, as time progresses, the construct retains its name but becomes based on a different set of initial and deduced hypotheses. For example in J. J. Thomson's time the hypothetical construct 'electron' was defined in terms of Maxwell's and Newton's theories (by combining equation (3.2.2), p. 82, with equation (3.2.3) one could obtain the equation of motion of an electron in a specified electromagnetic field). Later Bohr's model redefined the concept 'electron' by using the original theory (in the Hamilton–Jacobi version) jointly with supplementary quantisation rules (*cf.* W. Wilson, 1940, vol. 3). Subsequently Schrödinger's equation, together with the probabilistic interpretations of quantum mechanics, provided yet another redefinition, and then came Dirac's (*cf.* 1958) relativistic equation of the electron (together with its deductions and interpretations). And this process may go on, so that the hypothetical construct 'electron' may mean something different to people using different theories.

Fröhlich (1969), while reviewing Popper's (1967) paper, pointed out that the hypothetical construct of 'elementary particle' requires complete reassessment, since under specific conditions any elementary particle can be 'annihilated' by combining with an appropriate antiparticle (*cf.* Lüders, 1961), giving rise to another type of particle (for example, when an electron and positron interact to give rise to a photon). Again, certain types of elementary particles (such as photons) can be converted into a particle and antiparticle ('pair creation'). There exist also numerous decay relations for fundamental particles (baryons, mesons, leptons)—see Schweber (1964), Visconti (1969, p. 134). This raises serious problems for those philosophers who believe that elementary particles (which are hypothetical constructs) have reality status. The interconvertibility of elementary particles, postulated by theory (and backed up by a great deal of experimental evidence) poses for realists new questions of 'existence'. (Yet while different in detail these questions are no different in essence from those which arise in the dissociation theory of chemistry, where a complex molecule also represents something different from the components into which it can dissociate.)

Some psychologists (such as Hull) have tended to confuse unobservable variables of theories and hypothetical constructs. Deutsch (1960, pp. 4–5) mentioned that Hull referred to molecules, atoms, protons and electrons as intervening variables (that is, unobservable variables), although these are all hypothetical constructs. Thus, Hull appeared to have misunderstood the functioning of generic concepts like 'electron' in theoretical physics. Moreover, hypothetical constructs of certain theories may in some cases be observable (for example, a complete brain) without instruments, so that electrons and brains, while hypothetical constructs in physics and molar behaviour theory, respectively, *have different observational accessibility*. Deutsch (1960, p. 9), though critical of McCorquodale and Meehl's definition of a 'hypothetical construct', did not himself clarify the essential nature of a hypothetical construct (as a set of hypotheses, and so on). This is shown by his assertion (Deutsch, 1960, p. 10) 'that a molecule in the kinetic theory of gases is . . . the constituent of a system . . .'. Though this is perfectly correct, the fact remains that, in the *calculations* of the kinetic theory of gases, molecules do not enter as constituents or 'entities', but there occur only postulated equations of interactions, which follow the initial hypotheses concerning molecules of gases. (However, in some of the equations the *number* of postulated 'entities' per unit volume may enter.)

To sum up: terms like 'molecules', 'electrons', 'genes' or 'ribosomes' are hypothetical constructs which refer to a set of initial and deduced hypotheses and to accompanying semantic statements. This point of view is close to that adopted by Carnap (1956) (while Sellars (1948), Feigl (1950) and others preferred to view 'hypothetical constructs' as corresponding to real entities). I discussed the realist point of view earlier. At best realism is an extrascientific metaphysical belief. Its adherents could argue that, notwithstanding redefinitions of terms like 'electrons' in terms of repeatedly changing systems of hypotheses, the changing systems of hypotheses are only improved representations of an underlying reality. But one does not have to accept this point of view.

Alas, some people have mistakenly assumed that we can only formulate theoretical statements referring to hypothetical constructs, provided we *justify* first of all the 'reality status' of these constructs independently of *any* theory. However, as hypothetical constructs can only be defined *via* a set of hypothesised properties (or variables), such demands put the cart before the horse. It is also unwarranted, except as a metaphysical belief, to ascribe reality status to unobservable variables. For example G. B. Brown (1956), who appeared to be an adherent of 'naive realism' (*cf.* Feigl, 1950, for a discussion of this doctrine), believed, in common with many thinkers of the past, that scientists ought to describe the universe in terms of causal 'agents'. He assigned reality status to forces in classical mechanics and believed that such metaphysical interpretations are essential. However, we can safely reject this view, which adds nothing to the practice or the pragmatic value of science (see also Toulmin, 1953, § 4.4, for relevant discussions, and P.

Frank's (1957) evaluations of the metaphysical interpretations of physics then propounded by Margenau and Werkmeister).[5]

Some scientists and others whose ways of thinking favoured geometrical (that is, pictorial) representations generalised their idiosyncratic preferences into the general dogma that theories which are not *anschaulich* (formulated in terms of pictorial representations) are 'bad'. It is true that some theoretical statements gain a great deal (particularly in lectures) by geometrical representation, or may even be dependent on these (for example, the Feynman diagrams in quantum electrodynamics—*cf.* Schweber, 1964—or the Watson–Crick model of DNA). But this does not mean that hypothetical constructs or theoretical relations must in all cases be capable of geometrical illustration. Quite probably the brains of those scientists who think preferentially in geometrical terms were often guided to valuable ideas by first establishing geometrical representations of various hypotheses which were later translated into formal terms. Scientists, while engaged in theory construction, are permitted to let their imagination run freely and think of electrons, protons or ribosomes *as if* they were real and directly observable entities. However, this mode of brain representation must not misguide the possessor of the brain that his brain representations correspond necessarily to any existing entities, any more than brain representations of unicorns (by an artist who produces a drawing of such an imaginary creature) correspond to existing entities.

3.9 Causality and causal explanations

The notion of 'causality' has been extensively discussed by many philosophers and scientists (*cf.* Bunge, 1959). It seems at present difficult to formulate a universally applicable definition of causality, since different theories utilise causal relations in widely different forms (for example, causal explanations in molecular biology are mostly verbal, while those of classical mechanics are formal and mathematical). Some physicists, for instance, consider a *causal* explanation in classical mechanics as the determination for a system of its canonically conjugated coordinates ($q_i(t)$, for $i = 1, 2 \ldots n$) and momenta ($p_i(t)$, for $i = 1, 2 \ldots n$) as functions of the time, given their values at some initial instant of time $t = 0$. These variables have to satisfy Hamilton's equations of motion

$$\mathrm{d}p_i/\mathrm{d}t = -\partial H/\partial q_i \qquad \text{and} \qquad \mathrm{d}q_i/\mathrm{d}t = \partial H/\partial p_i \tag{3.9.1}$$

[5] As philosophers of science and scientists change their views not infrequently, the reader should bear in mind that opinions credited to various people at some particular time need not necessarily correspond to those held by them at present. I am fully aware of this, since my own views have undergone a radical change during the past ten years, as the result of much active theorising during that period.

where H is 'Hamilton's function' (that is, the 'Hamiltonian'; for a definition, see Goldstein, 1950) for a holonomic system.

In classical electrodynamics there are corresponding 'causal descriptions' resulting from solutions of Maxwell's field equations together with the constitutive equations for materials, subject to prescribed initial conditions and boundary conditions (*cf.* Stratton, 1941, and Born and Wolf, 1959, for applications to classical optics). (For statistical 'causal' relations between quantum electrodynamics and optics; see Glauber, 1970.) In quantum mechanics we have an equation of state in the form

$$H\psi = a\, \partial\psi/\partial t \qquad (3.9.2)$$

where H is the Hamiltonian operator of the system, a is a constant and ψ a 'state vector' in Hilbert space. Although time-dependent descriptions in quantum mechanics differ from those in classical mechanics, we can, nevertheless, determine the value of ψ at time t, given its value at time $t = 0$. We could say that mean values, and other statistical quantities, calculated at time t with the help of ψ, provide a *causal* description of the system. But we could abstain from this interpretation as it is simply a matter of verbal definition of causality.

Outside physics (for example, in many parts of molecular biology, and in some parts of the geological science of plate tectonics—*cf.* Alvarez, 1972, Le Pichon *et al.*, 1973—and other branches of science), some writers still use 'causal explanations' of the simple type: 'If event or state A takes place, then B is likely to *follow*'. Such verbal explanations remain useful in conditions where it is neither essential nor possible to formulate problems or hypotheses in terms of mathematical equations, nor to state initial conditions in precise mathematical terms. For example, the hypothesised explanation of polypeptide chain synthesis, starting with the formation of an 'initiation complex' which incorporates a particular type of methionine residue (*cf.* Wassermann, 1972), followed by *stepwise* incorporation of other specific amino acid residues into the nascent polypeptide chain, provides a typical example of verbal causal hypothetical event sequences, which have proved most useful in molecular biology.

Moyal (1949c, p. 310) emphasised that

> the concept of causal relations in the physical world is inseparably tied up with the concept of space and time: we cannot define the former clearly without having previously defined the latter. To take the simplest example, that of events occurring at a single point in space, we cannot say that event C *causes* event E if C and E are simultaneous; the idea of a causal relation between two events occurring in the same place can be conceived only if these events are separated by an interval of *time*.

In order to avoid an infinite regress, care must be taken to start a causal explanation from an assumed initial state of the system, and to suggest how subsequent events take place.

In § 3.6A, I have already mentioned Ryle's lack of appreciation that *scientific* (and in particular mechanistic) explanations have to start from an assumed initial state of a system. In his polemic (Ryle, 1949) against mechanistic theories of neuropsychology (which he headed the 'bogy of mechanism'), his misunderstanding of this essential point comes out clearly. He based his discussion on an anthropomorphic analogy. Suppose, he argued, we consider a game of any kind (such as chess) played by man. The game obeys rules, but which possible moves conforming to the rules are made by the players cannot be deduced from the rules but depend on the players' decisions. Hence, Ryle argued, the 'purpose' and tactics of the game are provided by something not inherent in the game. By *analogy* Ryle concluded that the laws of nature determine only how natural systems could behave, and that from the laws we cannot deduce in which situations these laws will be utilised in natural processes. While this is correct, it is apt to convey a wrong impression of the *aims* of mechanistic explanations, for the following reasons.

(i) Ryle invokes a dichotomy of game and tactician, assuming tacitly that the tactician's brain and motor system do not operate mechanistically. If the tactical decisions made by the players brains are *in their essentials* mechanistically explicable, then Ryle's dichotomous separation in the analogy is unjustified. It would be like saying that, even if we know the principles which explain how the earth's atmosphere would react to outside radiation, we cannot deduce from these principles what will actually happen when a flare erupts from the surface of the sun, as this depends on the behaviour of another system, namely, the sun. Ryle, like Polanyi (1968), disregards the fact that scientific idealisations often separate parts of coherent problems for individual treatment. When discussing the general effect of radiation on the earth's atmosphere, they may disregard the *sources* of this radiation but only assume, as an *initial condition*, that radiation is incident. (However, such idealisations may be dropped in some investigations; *cf. J. Geophys. Res.*, 1973, **78,** 6167).

(ii) The introduction of initial and boundary conditions into scientific problems is an abstraction, serving to disregard systems outside boundaries, and events which occurred before a certain selected initial time. This restriction, though essential for scientific analyses, provides no justification for the hypotheses of non-mechanistic 'agents' (such as 'purposes'—not as *defined* by certain mechanists—and other vitalist spooks). It implies, however, that the *scope* of scientific explanations is limited. Ryle might possibly agree that, although scientific explanations are restricted to idealised and generalised systems, non-scientific 'explanations' are invariably a concoction of empirically unsupported *ad hoc* hypotheses. Historians and lawyers may continue to 'explain' human actions in terms of 'purposes' and 'motives'. Such *ignotum per ignotius* 'explanations' may be satisfying to the man in the street, who is always a willing victim of word-substitution procedures, which pretend to 'explain' complex situations by cliché expressions.

Many people have failed to appreciate the limitations of scientific theorising. Theories always deal with *classes* of phenomena. They can only be applied *ad hoc* to explain unique and idiosyncratic behaviour of specific systems if it can be shown that such unique behaviour could be deduced from a theory by giving particular (*ad hoc*) values to some of its variables, and by choosing specific initial conditions and boundary conditions (where applicable). In the case of highly complex systems (such as man), theories of sufficient generality can only hope to explain features common to many samples of one class of system or to many different classes of systems. Without making a prohibitive nmber of *ad hoc* assumptions it seems unlikely that the idiosyncratic behaviour of particular complex organisms or of a specific man can be 'explained' in scientific terms. Science cannot explain why Napoleon or Cromwell made particular historically significant decisions. Scientists could discover that Napoleon and Cromwell conformed to certain behaviour patterns which are typical of certain classes of men, and they might one day explain the biochemical origin of such behaviour patterns. But beyond this one cannot go.

It appears that some philosophers have thoroughly misunderstood the fact that science is rarely if ever concerned with explaining idiosyncratic events, but aims to give explanations of properties common to many members of a class. Accordingly scientific explanations cannot be comprehensive or exhaustive, for the simple reason that their initial hypotheses idealise, and that their initial conditions cannot, in many, or all, cases take into account the past history of the system, which might be relevant (for example, specific engrams which a particular man's brain has formed as a result of years of perceptual learning). Scientists can explain general learning processes, but they are not concerned with the particular items which a particular man learns (unless they enter into the evaluation of a particular experiment, where it is not the items *per se* which matter but possible class characteristics of the items).

The following passage from Peters (1958, p. 12) illustrates the point of view of some philosophers:

> In the mechanical conception of 'cause' it is also demanded that there should be spatial and temporal continuity between the movements involved. Now the trouble about giving this sort of explanation of human actions is that we can never specify an action exhaustively in terms of movements of the body or within the body (Hamlyn, 1953). It is therefore impossible to state sufficient conditions in terms of antecedent movements which may vary concomitantly with subsequent movements. 'Signing a contract', for instance, is a typical example of a human action. The movements involved are grouped together, as I have shown, by means of norms governing the appropriateness of movements relative to a goal. But it would be impossible to stipulate exhaustively what the movements *must* be.

Peters then proceeds to cite examples of motor equivalence, for example that 'contract-signing' movements could be executed by many different motor systems of the human body. (For instance, a contract could be signed by holding a pen between our toes, or attaching it to a forearm, or inserting it into our mouth and making the appropriate head and body movements.)

Peters' argument thoroughly mistakes the aims of scientific explanation and research. Scientists are not concerned with providing *sufficient* (that is, complete) explanations of phenomena such as motor equivalence, but rather aim to suggest general hypothesised mechanisms which could explain all general major aspects of transfer phenomena, namely the processes whereby alternative muscle systems can be utilised in appropriate combinations so as to execute a specific 'planned' activity (such as writing one's signature). Scientific explanations start from initial assumptions, and hence do not aim to be exhaustive (as they would otherwise find an endless causal regress). Peters (1958) believed that psychologists ought to deal mainly with concrete (idiosyncratic or more or less unique) problems, concerning particular actions of particular people. But this is alien to the basic outlook of science, which is concerned with class properties. (See also Grant's (1955) discussion of Ryle's (1949) book.)

Many everyday events exhibit features which to some extent are amenable to scientific explanation (for example, when somebody suffers from the symptoms of a common type of bacterial infection, his symptoms can be partly explained in some cases). However, the 'depth' of explanation in such cases depends (a) on how far back in the causal chain one is prepared to proceed, and (b) on what type of theory one adopts.

Some scientists and philosophers have posed the question: is human behaviour causally determinate or indeterminate? As exhaustive causal explanations are impossible (see my reply to Peters), the issue must remain undecidable (see also § 4.3B). However, we can formulate different problems for the more restricted scientific approach to brain functions. Several scientists have assumed that brain mechanisms in neuropsychological processes can only be explained in terms of statistical theories, while others believe that the theories concerned can be formulated (notwithstanding hidden statistical aspects, which are present in any theory; see above) in non-statistical form (*cf.* Clarke, 1958; L. F. Jackson, 1958; Burt, 1958a, for different points of view). As brains consist ultimately of molecules, and as the description of the behaviour and structure of molecules at the level of quantum chemistry demands a statistical treatment, it would seem that *prima facie* a statistical theory is required. However, on the same grounds, molecular biology should have to be an essentially statistical theory. But this is not the case, because people operate in terms of larger units, which are already (implicitly) averaged out.

For instance, when Watson and Crick established their model of the DNA double helix, their wood and wire models of atoms and molecules completely disregarded statistical distributions of electrons within atoms and molecules, but only assumed (implicitly) that the statistical electron distributions (and protons present) cause the appearance of particular hydrogen bonds and bond angles. In fact, much chemical model building (including Pauling's discovery of the polypeptide α-helix conformation: *cf.* Wassermann, 1972) depends on such simplified assumptions, which consider molecules as systems whose internal statistical properties are not required for the purpose of model building.

Similarly, when developing neuropsychological models which operate at the molecular level, one can adopt a corresponding non-statistical strategy. (In fact, my recent theory of morphogenesis (Wassermann, 1972, 1973, 1974) and my new brain model (now being developed; see Wassermann, 1972, chapter 17, and 1974) are based throughout on assumed statistically averaged out properties of macromolecules.) The issue concerns simply the level of theorising which one wishes to adopt. If we assume, as far as theorising is concerned, that molecules have non-statistical properties such as directed bonds of specified lengths and bond angles, then we do not have to provide a quantum-mechanical explanation for these bonds at this level of theorising. These remarks apply to many types of theories which only consider a 'coarse' level of molecular structure in molecular biology. For instance, models of cell surface membranes or theories of protein composition of ribosomes (*cf.* Wassermann, 1972, for detailed discussions) do not have to invoke quantum mechanics in order to understand essential features. However, quantum chemistry has proved useful in suggesting interpretations of detailed features of specific macromolecules (Pullman and Pullman, 1963).

Those who mistakenly believe that in theorising about brains and organisms one must invariably resort to the most complex level of theorising, namely quantum mechanics, have produced some insupportable arguments. For instance, Clarke (1958) cited Broad's (1925, p. 435) view that

> If the Quantum Theory be correct, we are probably witnessing a breakdown of the accepted assumption . . . (to cite Clarke, 1958) the assumption, namely that the 'minute structure' of the brain as of other physical substances, is subject to the same laws as the gross 'structures' which we actually observe.

In view of my preceding remarks, it is seen that Clarke's conclusions need only be relevant if, in theorising about brains, we have to invoke quantum mechanics. But for many brain processes there is, in all probability, no need to invoke quantum mechanics *explicitly*. The erroneous view that quantum mechanics must be invoked explicitly in cases where there appears to be no need (at least at present) was defended by the late Sir Cyril Burt (1958a, p. 77). He cited Bohr, Schrödinger and others as having pointed out that 'the gene and the neural synapse are sufficiently small to make it probable that their essential functions ought to be conceived in terms of the quantum theory rather than the classical theory.' Unfortunately Burt did not seem to know (a) that the Watson–Crick theory does not (explicitly) depend on quantum mechanics, and (b) that a synapse may contain thousands of macromolecules whose arrangements could in sufficient approximation be described without resort to quantum mechanics. In fact, X-ray diffraction theory combined with model building has in many cases allowed structures to be elucidated without resort to quantum mechanics (for example, in the case of DNA, the α-helix structure and several proteins—*cf.* Wassermann, 1972).

Yourgrau (1970, p. 125) has already noted that 'the [quantum-mechanical] uncertainty principle is still subject to many inane interpretations, so many indeed that we can ascribe practically any metaphysical nonsense whatsoever to it'. In this context let us briefly examine Burt's (1961, p. 60, 1962) *a priori* assertion that brains must function probabilistically. To substantiate his argument he applied the uncertainty principle in a physically impermissible way to a system as massive as a synapse. Referring to an 'easy calculation' (Burt, 1961, p. 60) which he never gave explicitly, and citing the supposed weight of the synapse *only*, Burt concluded that there must be an uncertainty of position of a synapse of about 1 Å. Yet, in the case considered, Heisenberg's uncertainty principle (*cf.* Born, 1948, appendix 22) refers to an inequality involving *momentum* and position of a point particle. Since Burt did not cite the totally unknown, but implied, momenta of synapses, it would seem that he may have mistaken weight for momentum. I conclude that Burt's quantum-mechanical arguments that brain function must be treated probabilistically rest on false premises.

3.10 Models as representatives of theories, and 'general systems theory'

The word 'model' has numerous connotations (*cf.* Byerly, 1969). For some people the dividing line between the words 'model' and 'theory' is thin, and possibly many theorists use the words interchangeably, as if they were synonymous (*cf.* Achinstein, 1965, p. 103). Others, however, consider the word 'model' as referring to a simulator system constructed of appropriate hardware (for example, the Watson–Crick 'model' of a DNA double helix) which represents in terms of its components some or all salient features of a theory, previously only worked out on paper. I shall discuss several meanings and uses of models.

(i) A simulator system (for example, a computer program serving as a simulator model) may be able to show up features of a system which may be difficult to deduce by formal mathematical procedures or by ordinary reasoning from the initial or deduced hypotheses of a theory. (Hesse (1953, 1964, 1966) and Braithwaite (1953) gave general discussions of models and pointed out inherent dangers of their use.)

(ii) Many physicists, social scientists (and so on) refer to a simplified version of a theory, which exists also in more highly developed form, as a 'model' (for example, the 'model' of a gas in which 'hard' (inelastic) spheres are substituted in place of elastic, non-spherical molecules). In this case the word 'model' implies additional idealisations not present in the fully formulated theory. Such additional idealisations may be important in cases where the originally richer theory is too difficult to handle.

(iii) Sometimes people will use a 'model' (for example, a hardware model, or an idealised form of a theory) simply to represent some specific salient features of a system, without any claim that their model is related to the actual mechanisms of the system concerned. For instance many (circardian) biological clock models have been proposed (*cf.* Pavlidis, 1971; Pittendrigh, 1961) which simulate observed features of biological clock *behaviour*, without claiming to simulate or represent the clock mechanisms.

(iv) Occasionally, as is well known to 'general system theorists', the initial and deduced hypotheses of a theory may contain variables or hypothesised properties which, if semantically reinterpreted, can serve as generators of structurally isomorphic theories, which are being applied to systems with totally different semantic contents. For example, electric network theory can be used as a formal instrument for simulating the behaviour of certain acoustical systems (*cf.* Morse, 1969) by representing specific acoustical components by presumptive functionally equivalent combinations of electric components. Terms which in the electric network are referred to as capacitances, inductances and resistances are being given appropriate reinterpretations in terms of acoustic components. Von Bertalanffy (1950a, b), in his 'general system theory', was one of the first to appreciate in full generality that *formally* identical but empirically different theories could be generated in different fields of science by giving different semantic interpretations to the terms which appear in a theory. However, system simulation was practised (in acoustics and elsewhere) long before Bertalanffy's formal discussions.

In some cases a theory postulates *specific structures* as well as *functional properties* of these structures. An adequate model representation of the theory demands then that the model represent the theory's structures as well as their functions. Partial simulation by a model can be misleading, *a point often disregarded by brain-model designers*. If a brain-model is intended to relate empirical generalisations about human or animal behaviour with postulated functions of internal and surface structures of neurons, then we must use a behaviour simulator model, whose structural constituents also simulate the *postulated* structures of the nervous system. In other words, if a simulator system is to reproduce relevant behaviour of a biological system, it must do so in terms of simulator components which resemble the components of the biological system as far as *all* postulated properties of the latter are concerned. Moreover, the simulator system must not introduce additional properties of its own which are not present in the system which is being simulated and which could significantly alter the results to be expected.

I conclude that simulation by models (as distinct from direct theorising) is potentially a very dangerous practice, which could introduce artefacts of an unforeseeable nature.

3.11 'Confirmation theory' for theories, and 'explanatory power'

Much effort has gone into attempts to discover 'measures of confirmation' or of 'explanatory power' of parts or whole of a theory (*cf.* Carnap, 1956; Scheffler, 1963; Hempel, 1966, 1967; Suppes, 1966; Rozeboom, 1968, 1971; Buck and Cohen, 1971; Jeffrey, 1971; and the symposium edited by Lakatos, 1968). 'Explanatory power' can be defined in various ways (see alternative definitions provided by various philosophers of science, and discussed by Jeffrey, 1971). Jeffrey (1971) considers 'the idea of using information-theoretical concepts to quantify the notion of explanatory power of theories' an attractive one. He believes that 'where the theory is adequately represented by a probability measure p on a Boolean algebra A of propositions, it is natural to think of measuring explanatory power by some function $H(A)$, the *entropy* or *uncertainty* of p'. Assuming that the Boolean algebra A is finite, and letting 'a' range over the strongest propositions in A which are not logically false, Jeffrey proposes as a measure of explanatory power the information function

$$H(A) = -\sum_{a} p(a) \log p(a) \tag{3.11.1}$$

where we set $p(a) \log p(a) = 0$ in the case $p(a) = 0$. According to Jeffrey, '$H(A)$ is greater the more ignorant we are about A'.

There are two aspects of confirmation theory which require attention. First of all, attempts to construct probability measures of confirmation which are based on the calculus of formal logic lead to serious difficulties which appear irremovable (*cf.* Rozeboom, 1968). Secondly, and this point is far more important, it is widely taken for granted by philosophers of science that many deductive propositions of scientific theories are equivalent to the use of formal (that is, mathematical) logic based, say, on Boolean algebra. However, this is not the case since, as repeatedly stressed, much of scientific deduction relies on approximation procedures, and computations with possible accumulative errors and so on. (In fact, the use of any mathematical formula which cannot be precisely satisfied by rational numbers involves approximations—even the evaluations of expressions of simple functions like logs, sines and cosines must terminate in approximations after a finite number of steps.)

Hence, to base calculi of confirmation or 'explanatory power' on probabilities which are related to propositions of Boolean algebra is an enterprise doomed to failure in most cases, with few (if any) exceptions. (Some important exactly soluble problems in classical and quantum mechanics are partial exceptions—such as the solutions of Schrödinger's and Dirac's equations for the hydrogen atom.) We could hardly allocate any probabilities to initial hypotheses of a theory if these are indirectly tested *via* a long chain of deductive intermediate hypotheses, which involve major mathematical approximation procedures, since we cannot know how these approximations (which cannot be expressed in terms of Boolean algebra) affect the probabilities. Hence, independently of Rozeboom's arguments regarding pitfalls in existing 'confirmation theory', there exist in many parts of science strong practical reasons, disregarded by the abstract formulations of 'confirmation theory' and 'explanatory power', suggesting that these notions are unlikely to be applicable in practice.

There is also the point, made by Rozeboom, that 'non-truth-functional connectives such as "is a cause of" are essential for expressing the laws of a world which is truly lawful, even if only probabilistically so'. (It must be stressed that many connectives in more complex science are not of the simple type 'is a cause of'. Yet, when more complex connectives are used, for instance in dynamics or quantum mechanics, we have, more often than not, confirmational uncertainty created by approximations and the like.)

Some scientists have argued that theories should only postulate 'testable hypotheses'. By this is meant no more than that the initial hypotheses of a theory must, *via* intermediate hypotheses, be related to empirical generalisations. It is not intended to convey direct testability of initial hypotheses, although some unenlightened people have misconstrued the word 'testability' and applied it to initial hypotheses. Those who make such statements invariably mistake initial hypotheses for empirical generalisations and are apt to confuse facts and hypotheses. (For instance they argue that certain spots seen in electronmicrographs, which are postulated to be ribosomes, *are* ('real') ribosomes.)

Ultimately, as stressed earlier, only empirical generalisations can be directly 'confirmed' (subject to acceptance of arbitrary fitting methods such as least-square fits). Consequently no *past data* can be verified by future ones. Only hypotheses (empirical generalisations *extrapolated* from past observations) can be confirmed (within the limits of accepted accuracy) by future observations. Statisticians have evolved rules for deciding when certain empirical generalisations may be rejected (*cf.* Wald, 1952). These so-called 'significance tests' rely on conventions, that is, social criteria of acceptance. As they cannot logically be justified, we cannot be certain that the same conventional standards are *a priori* appropriate for all branches of science. Accordingly (see also pp. 67–68), some scientists have felt hesitant to accept empirical generalisations on the strength of significance tests only. I pointed out that linkage of empirical generalisations *via* a network of hypotheses of a theory may help to remove some (but not necessarily all) of the doubts concerning any one empirical generalisation. For instance, major earthquakes are relatively rare events. By explaining earthquakes in terms of the dynamics of geological plate tectonics, one can deduce that earthquakes are likely to occur along detectable geological 'faults'. Hence if it can be shown that a place where an earthquake reported centuries ago is supposed to have occurred lies in the vicinity of a known fault, this increases the plausibility of the report. (The report as such is equivalent to a hypothesis.)

The inappropriateness of treating scientific hypotheses as if they were purely logical propositions (and of associating these with 'confirmation measures') is also suggested by Goodman's 'green–grue paradox' which has aroused interest among some philosophers of science. Hesse (1969) has described this alleged paradox as follows:

> 'Grue' applies to all things examined before [time] *T* just in case they are green, but to other things just in case they are blue (Goodman, 1965, p. 74). Now consider the two hypotheses 'All emeralds are green', 'All emeralds are grue'. Each is supported *by the same evidence* before *T*, for all emeralds observed have in fact been green before *T* and therefore grue. By induction by simple enumeration of positive instances, both hypotheses are equally supported. But they yield different contradictory predictions after *T*, and we should have no hesitation in accepting the prediction of the first rather than the second. Therefore in any adequate inductive theory they ought not to be regarded as equally supported. In Goodman's terminology, we ought to be able to show that 'green' is more *projectible* than 'grue'. (Some italics are mine.)

The preceding argument (i) omits precise statements of the (assumed) 'evidence'. Without precise accounts of evidence, scientists cannot assess the *claim* that this evidence genuinely applied to all samples. Hesse presents a generalisation ('applies to *all* things') as if it were factual evidence. (ii) The argument also assumes that *particular* evidence can be regarded as sufficient. Yet we can never know whether we possess *all* the relevant evidence sufficient for assessing a state of affairs. We can at best only *believe* that we have encompassed the relevant evidence. In order to obtain evidence, certain empirical tests or observations may be required (such as spectroscopic tests), and we simply cannot know how many different kinds of tests may be required at any one time to say with any measure of assurance that two different samples (say of an emerald) will continue to have the same properties.

Systems which often have concomitant properties may have one but not the other of these properties abolished by experimental procedures or naturally occurring events. Hence statements of the type '*All* things which have property A also have property B' are simply not scientific inductions which can claim to be valid *for all time* because A and B may be dissociable. Some philosophers of science (Goodman, 1965, and others) have not attached enough importance to this dissociability. Strain-specific bacteria can suddenly undergo mutations, retaining property A (previously possessed) but losing property B. For example, suppose *E. coli* cells of strain *X* share in most cases properties A and B. But suppose that mutants of strain *X* (which appear after time *T*, say) retain property A, but not property B. If property A in the absence of property B allows *E. coli* cells still to be characterised as being of strain *X*, then absence of property B does not affect the usual strain classification. But the *evidence*—that is, the tests—for properties A and B may differ.

The fact is that we can never assert with certainty that any property applies to *all* things supposedly of class A at specific times, as many systems are of immensely complex internal constitution, allowing for unobservable internal variations of samples (mutations of bacteria, unsuspected or uncontrollable impurities of samples, and so on). Any system can, through unknown external or internal causes, change some of its components and hence one, but not necessarily others, of its properties (for example, oxidation or random radiation could induce mutations; or consider induced dislocations in metals). Empirical generalisations relating to systems can therefore never state something which applies *necessarily* to *all* members of a class at specified times.

In the case of the emeralds, as in *E. coli* of strain *X*, green and property A may respectively be major identifying properties of the objects concerned, while this does not apply to 'grue' and property B respectively. If green is a 'coarse' major identification property of emeralds (and crystal structure a finer feature only detectable with instruments), then we would not readily call a yellow gem with the crystal structure of an 'emerald' an emerald without, say, X-ray crystallographic tests, and reasons for showing that it has many other major properties (such as composition or structure) of emeralds in closely similar form. Thus, whereas 'green' and a certain outward shape may be major identifying properties of emeralds, 'grue' could be a minor property. Hesse's account, for instance, gives no indication as to what role 'grue' plays in the normal identification (and hence classification) of emeralds. She correctly noticed that there exists an asymmetry as to why green is more 'projectible', but does not appear to state that this asymmetry could be connected with distinctions between major and minor identificational (or classificatory) properties. The reason why we classify *E. coli* cells as *E. coli* cells (even those of many different strains) is that they contain major recognisable properties common to all strains; otherwise we could not classify these cells.

There is also the doubtful assertion (in Hesse's account of Goodman's paradox) that two different properties are supported by the same evidence. If the presence of the two properties were completely based on the same evidence, it would be difficult to know how their independent presence could be asserted. Hence the methods of observation with respect to the two properties *must* differ. All that could be claimed is that the different evidence for both properties appeared concomitantly up to time *T*. But different evidence may be linked to different internal mechanisms which, while often linked, could be decoupled in other cases. The paradox arises from talking loosely about 'evidence' without making it clear that the evidence for different properties (such as green and grue) must be based on different observation procedures, as otherwise one could only sample for one and not two different properties. (Feyerabend (1968) thought that Goodman's paradox provides a clear refutation of the idea that a hypothesis can be confirmed—see also Foster's (1969) counter-argument. As examples of *E. coli* mutants show, Feyerabend would only have a case if a deduction from an initial hypothesis of a theory implied that a deduced empirical generalisation *must* hold *eternally*. However, as no initial hypothesis of a theory is (or can be) assumed to hold for all times, eternal validity of deduced EGs could never be entailed, even where strict logical deductions are feasible. But this does not exclude confirmability of hypotheses for *long periods*).

3.12 Theories as instruments of prediction

The technological applicability of theories has misled some people into believing that the 'sole purpose of scientific theory is to predict the most likely course of events and to control in some measure the events themselves' (Rawcliffe, 1952, p. 489). This point of view completely disregards the explanatory and ordering function of theories. Predictive capacity, when present, is an important and attractive additional property of theories, but is not essential as far as explanations of already established empirical generalisations are concerned, and has no bearing on the goodness of fit of already established empirical generalisations with respect to previously observed data. Only if it is claimed that the empirical generalisations are to be valid for an indefinite (but not infinite) period must further observations, relating to these generalisations, then agree within the accepted limits of confidence with the already established empirical generalisations, or otherwise be considered as casting doubt on the latter.

The theory of evolution, for instance, is not intended as a predictive tool (Scriven, 1959), as it does not claim that any specific evolutionary trend will persist, or suggest which new trends will appear. The concern is with *mechanisms* of evolution, which are of a general kind. However, general mechanisms can only be applied to *specific* systems if the parameters which may appear in the general theories (or other particular conditions such as initial and boundary conditions) can be adequately hypothesised for these systems. This is not possible for the theory of evolution since, for instance, we do not know the complete chromosomal DNA constitutions of individual members of a species and hence the differential susceptibilities of different DNA stretches to random radiation or other natural mutagenic agents. Nor could we at present predict how precisely such mutations would affect the phenotype. (Environmental effects on evolution are equally difficult to predict.)

Explanations have almost invariably preceded predictions. Newtonian classical mechanics was originally developed to *explain* Kepler's *already known* laws and other *known* empirical generalisations, and it was one of its earliest triumphs that it succeeded in this. Likewise, electrodynamics was developed to *explain* the already known empirical generalisations of Faraday and others, and much of quantum mechanics served primarily to create order among already known phenomena by *explaining* numerous empirical generalisations (for example, of spectroscopy). Superconductivity was only explained but not predicted by quantum mechanics (see Harrison, 1970). The structure of the emission spectrum of hydrogen atoms was known long before quantum mechanics explained it, and the Lamb shift was known before quantum electrodynamics accounted for it. Quantum chemistry is still struggling to explain many known facts rather than to predict new ones.

Nevertheless, electrodynamics has been extensively used as a predictive tool in electrotechnology. Likewise, classical mechanics, aerodynamics, and so on, have had innumerable industrial applications relying on predictions. Continuum mechanics provided much help in predicting (approximately) the properties of structures (*via* the linear theory of elasticity), and the theory of vibrating solids has allowed numerous predictions useful in the design of acoustical transducers (microphones, loudspeakers, and so on). Many other important applications of theories (as of hydrodynamics to ship design) could be listed. Hertz's 'prediction' of electromagnetic radiation (see Whittaker, 1951, pp. 324 ff. for a historical account), as a deduction from Maxwell's theory, with its ultimate consequences for radio and television inventions, and the design of many optical instruments which rely on electromagnetic theory (or its approximate version, ray optics—*cf.* Born and Wolf, 1959), show the great predictive and pragmatic value of such theories.

Predictions by a theory can be of at least three types:

(1) Already *established* and well confirmed empirical generalisations may be used to make predictions about the outcome of future experiments in relation to these generalisations: for example, an appropriate equation of state may be applied to a simple gas to predict its hypothetical temperature–pressure–volume dependence. In fact when, say, chemists or molecular biologists describe a specific, previously used, method, they expect ('predict') that it will produce the same 'effects'.

(2) Previously unknown and untested empirical generalisations are sometimes deduced from a theory, leading thereby to discoveries of new types of phenomena or (hypothetical) systems. Dirac's prediction of the positron on the basis of his linearised relativistic equations for the electron and positron, Josephson's prediction of supercurrent tunnelling (*cf.*

Harrison, 1970, p. 516, for a review) with conservation of the number of particle excitations, electromagnetic radiation as a deduction from Maxwell's theory, the discovery of an elementary particle (a neutral kaon) on the basis of the Gell-Mann theory (*cf.* Gell-Mann and Pais, 1955, and discussions in Schweber, 1964, p. 288), provide typical examples. Unconfirmable predictions need not necessarily invalidate most of a theory. In particular, if the prediction involves inadequate approximation procedures, discrepancies could be caused by the latter.

(3) As mentioned, some theories contain *systemic parameters* in, say, the main equations or boundary conditions (for instance, the refractive index in Snell's law, diffusion coefficients in the linear partial differential equations for diffusion: *cf.*, for example, Wassermann, 1952). These systemic parameters, while assumed to be constant for each system of the class considered, may take different values for different systems of the same class. Linear electric network theory in its most general form (*cf.* Jaeger, 1949, p. 32) postulates circuit relations in terms of inductances, capacitances, resistances and mutual inductances, which form systemic parameters. By substituting different values for these parameters, the properties (such as time-dependent behaviour) of different networks can be deduced or 'predicted'.

A theory makes predictions of type (1) provided that its empirical generalisations are assumed to be of some durability. Type (2) predictions are not essential for the explanatory validity of a theory. But a type (2) prediction may add to the prestige value of a theory. If a significant and confirmable type (2) prediction can be made, it strengthens our confidence in the theory (H. Meyer, 1951), while an unconfirmable prediction might create doubts. In fact, if Dirac's theory had not predicted the positron, its statements concerning electrons would have been as valid as they are now, except that interpretative difficulties might have occurred in connection with processes of electron–positron pair production and annihilation.

Factual incompatibility with a newly predicted empirical generalisation does not invalidate a theory as a means of deducing and linking those empirical generalisations which agree with it. It does, however, limit the scope of the theory. Thus, when classical mechanics was found to be inapplicable to domains of atomic or subatomic dimensions, this did not diminish its applicability to the motion of systems of sufficiently large dimensions. Engineers do not require quantum mechanics in order to design bridges or automobiles.

Type (2) predictions may be 'fruitful' and stimulating by suggesting new experiments, thereby keeping scientists occupied. The raising of new problems by the solution of existing ones is, in fact, desirable, as otherwise scientific research would come to a standstill. But it is not logically essential. The faulty impression persists that type (2) predictions are essential for the validity of a theory. However, whether a confirmed empirical generalisation

(EG) is deduced from a theory after discovery of this EG, or 'predicted' from a theory, and confirmed subsequently, is purely a matter of historical accident (*cf.* Meyer, 1951). It is quite irrelevant to the validity of the theory and its network structure. Confirmation, as well as contradiction, of a predicted empirical generalisation is certainly of significance for the status of the theory (for example, by increasing faith in, or by discovering limitations of, the theory; see above).

Many distinguished experimenters and theorists have failed to appreciate the explanatory function of theories, and have given undue weight to their predictive aspects. This is not surprising because of the prestige implications of correct predictions and their possible technological implications. Predictions can be of greater *social* benefit than explanations. Indeed, engineering helps man's daily needs more than Beethoven's string quartets or Shakespeare's plays. But scientific *explanations* add to man's understanding of his environment which many scientists cherish as much as they enjoy Beethoven's or Shakespeare's works. Radioastronomy helps us to explain the structure of the universe, thereby satisfying our curiosity; but it is not intended to improve our everyday living standards. Man does not live by bread alone.

Now for some examples of unjustified claims. Eysenck (1953, p. 235) argued that it is 'not *ex post facto* explanation which constitutes science, but prediction which can be verified'. One wonders whether Eysenck has studied the history of quantum mechanics or quantum chemistry, or of Newtonian mechanics. Darwin's theory of evolution[6] also provided *ex post facto* explanations, as did theories of plate tectonics and of astronomy

[6] Lewontin (1972) discussed the application of Popper's philosophy to the theory of evolution. While, in the opinion of Popper and others, evolution (like history) represents sequences of unique events, many evolutionists (such as E. Mayr) recognise class characteristics in evolutionary processes. As Lewontin put it: 'uniqueness is in the eye of the beholder. The reason that Popper does not reject the rotation of the Earth on its axis as an unfit subject for science is that, from the standpoint of the celestial mechanic, one day is like another, even if that is not true for the journalist.'

Lewontin believes that the non-falsifiability of postulated natural selection in eras long past makes natural selection a metaphysical hypothesis, from Popper's point of view. Lewontin argues that evolutionists, instead of accepting the falsification criterion, rely on the confirmation of natural selection by experimental procedures, which deal with specifically created conditions. These, of course, can only confirm that natural selection of the type observed in specific situations *could* have been operative in evolution, and more direct confirmation for this is always welcome. Admittedly, one can invent *ad hoc* hypotheses as to how natural selection could *in the past* have acted in specific cases, and contemporary experiments are no safeguard that similar factors acted in the past. However, *class characteristics* of evolutionary processes could help to eliminate most or all *ad hoc* hypotheses. (The reader is referred to Lewontin (1972) for a brief explanation of current trends in experimentation, and the important distinction between 'tautological' and 'functional' selection.) I agree with Lewontin's statement that 'Natural selection of the character states themselves is the essence of Darwinism. All else is molecular biology.' In fact it is the task of molecular biology to explain how genetic changes could have led to the production of new and advantageous characters, as natural selection by itself does not create organismic novelty (see Wassermann, 1974, appendix 1, for novel molecular biological explanations which seem to harmonise with some of E. Mayr's views).

(*cf.* Scriven, 1959; Mayr, 1961). Woodworth and Schlosberg (1955, in the caption to their fig. 14.17, p. 418) claimed that 'a crucial test of any theory is the ability it confers not so much to explain facts already known as to predict the results that will occur under certain untried conditions'. Again, G. Murphy (1953, p. 47) believed that 'a really good theory is a theory which *immediately suggests empirical investigations* which will tend either to confirm or repudiate it'. Judged in this way, much of quantum mechanics is not a really good theory, unless one regards repetition of the same *type* of experiment as crucial. (Likewise, C. L. Hull did not appreciate that the capacity of a theory to make a type (2) prediction is irrelevant to its intrinsic value—*cf.* Koch, 1954, p. 12 (b).) Another example of a similar view is provided by G. B. Brown's (1952, p. 134) arguments.

Additional examples illustrating these widespread misunderstandings are also cited by Scriven (1959), whose important paper on this topic relates closely to the preceding discussions. Some writers, unfortunately, use the word 'prediction' in the sense of *making further deductions* from theories whenever possible. This, in any case, is an obligatory task for theoreticians. Marx (1952) aptly quoted Brown's (1936) dictum that 'inability to predict or control earthquakes [at that time] has not prevented the significant scientific development of seismology. (But see Lomnitz, 1973.)

Sometimes the systemic parameters may vary in a manner which cannot be fixed by or be known to an experimenter, thus giving rise to the fickle behaviour described by Hebb (p. 65 above). But we must not be rash in ascribing, without good reasons, any deviations from known empirical generalisations to uncontrollable variations of systemic parameters. Yet the difficulties which many behavioural scientists experience in repeating each other's results in human and animal experiments suggest that the uncontrollable variation of systemic parameters in these *self-organising systems* may play a significant role, in contrast to most of the systems studied by physicists, where strong 'fickleness' of behaviour is a rare phenomenon. (Brownian motion is not comparable!)

It is surprising to see some psychologists, and even some biologists, complain at times that they failed to obtain consistent results *indefinitely* by experimenting with the same organism (for example, a human subject), or that similar results could not be obtained when experimenting repeatedly with organisms of the same species. Surely it is absurd to assume that the internal parameters of self-organising, complex, multi-parameter organisms should retain constant values over long periods. Memory retrieval systems and internal 'sets' (in the sense of behaviour-determining tendencies) could change slowly or abruptly (see Hebb, cited on p. 65). It is gratifying that under strictly stimulus-bound conditions (as in classical conditioning experiments, or in instrumental conditioning) experimentalists have often obtained approximately repeatable results with some organisms of the same species.

But to reverse the argument and to demand that the behaviour of highly flexible organisms should correspond to the stereotyped behaviour characteristic of inflexible organisms (*cf.* some typical examples studied by ethologists: *e.g.* Tinbergen, 1951; Hinde, 1966; Lorenz, 1966) is a request that certain organisms should not behave as they do. I believe therefore that those who assert that the discovery of 'strictly repeatable' experiments in psychology or biology is essential, are in search of a mare's nest. I shall return to this topic in the following section.

It has been argued that '. . . unpredictability in principle has always been considered as the salient point of emergence . . .' (Popper, 1970, p. 12). The preceding discussion shows that this does not follow, since the degree of predictability is related to controllability and complexity of systemic parameters. Philosophical 'thought experiments' about what is possible 'in principle' for complex organisms are useless, because in practice these organisms cannot in certain experiments be rigorously controlled. I consider it therefore inappropriate to base 'emergence' on criteria of unpredictability.

3.13 Types of repeatability

Each empirical generalisation demands a type of repeatability *appropriate to its claims*. (I owe this remark to Professor R. Harrop; *cf.* Wassermann (1955) for a more detailed discussion.) The degree of repeatability has only to match the observed frequency of the phenomena as generalised by appropriate probabilities (or expectation values, as the case may be). Consequently rare phenomena (such as specified genetic freaks, specific volcanic eruptions or major earthquakes) do not require repeatability on demand for their confirmation. Alas, numerous physics and chemistry experiments done in schools (where nobody has time to study rare phenomena) give many people already in their youth the misleading impression that the type of repeatability of these simple experiments must form the prototype for all scientific work. For instance Fraser Nicol (1955, p. 72) assumed that 'repeatability' means

> the designing of an experiment which, found in practice to produce a significant effect, can be repeated by any competent person at any time in the foreseeable future with approximately similar significant results.

Repeatability on demand is certainly excellent if attainable. But major earthquakes, unexpected discoveries of supernovae, and other phenomena not reproducible on demand (particularly some relating to organisms), suggest the futility of requesting that a type of repeatability appropriate to one class of systems or events should be universally applicable to all classes of systems or event. Those who think otherwise should be asked to produce five hundred living composers of Beethoven's calibre on demand.

4 SOME REMARKS ON THEORETICAL APPROACHES TO THE LIFE SCIENCES

4.1 Intervening variables and empirical anchorage

Although theoretical formulations of learning, motivation and other branches of behaviour science have nowadays shifted predominantly to statistical models, which in most cases attempt no more than to provide an *empirical generalisation* regarding a specific form of behaviour of a particular species, older attempts tried something more radical. The *molar* behaviour theories of (a) Hull (1943, 1950, 1951, 1952; Hull *et al.*, 1940; continued by Spence, 1956) and (b) Tolman (1932, 1945, 1949, 1955) are two of the most prominent systems of the older type which attempted to make use of intervening variables (see § 3.5). I have mentioned that molar behaviourists consider brains as hypothetical constructs (p. 104). The work of Hull and Tolman was critically surveyed in a book edited by Estes *et al.* (1954). Koch (1954), in a thorough critical survey of Hull's theoretical formulations (see also Cotton (1955) and Seward (1954) for additional criticisms of Hull's system), concluded that Hull failed to establish his theory 'because he did not adequately meet concrete problems of *empirical definition*, of *measurement*, or *quantification*, of *intervening variable construction* and of various sub-specification of all of these'.

More explicitly Koch concluded that:

(1) 'Secure anchorage, either in a quantitative or qualitative sense, does not hold in a single case for the relations of systematic independent and dependent variables to their intended range of reductive symptoms.'

(2) 'No given intervening variable is securely and univocally anchored to its relevant systematic independent and/or dependent variables, either quantitatively or qualitatively.'

(3) 'No given intervening variable is related to any other intervening variable in the chain with sufficient determinacy to permit quantitative passage from one to the other, nor are certain of the variables, and the relations connecting them, defined with sufficient precision to permit "qualitative" passage.'

Tolman's system was formalised by McCorquodale and Meehl (1954), to whose discussion and evaluation the reader is referred. Whereas Hull's theory aimed at quantitative stimulus–response relations, Tolman's system, if further developed, could presumably be formulated in terms of probability matrices. His concept of 'expectancy' with its implied selectivity at multiple choice points might favour such an interpretation. The molar approach of Tolman and Hull, and of those later theorists who relied on stochastic models of molar behaviour, represent typical examples of 'black boxism'. Hull tried to imitate the older and then better established sciences, by introducing intervening variables (that is, unobservable variables; see § 3.5) in attempts to link up empirical generalisations *via* initial hypotheses. However, in common with many other molar behaviour theorists, he did not appear to appreciate one important aspect of many advanced theories, namely that they explain molar properties in terms of detailed hypothesised internal structural components of the system considered. For man and animals, some of the most relevant components involved in behaviour are the constituents of the nervous systems of the organisms concerned.

One could expect that, if stimulus and response variables can be related at all *via* a theory, then they must be related *via* the central nervous system. Probably the fact that unconditioned reflexes (though also under partial central control) are automatic, and that conditioned reflexes are related to unconditioned reflexes, may have misled many earlier psychologists into thinking that the nervous system operates in a very simple manner. However, such ideas are quite erroneous. Behaviour, as a rule, cannot be explained simply in terms of sequences or hierarchies of conditioned reflexes, and 'expectancies' (and the like) must be allocated hypothesised brain function correlates.

4.2 Psychoanalysis and theory construction

Several philosophers of science (*cf.* Bunge, 1967c, vol. 1, and the symposium edited by Hook, 1959) and scientists (such as Eysenck, 1960, 1961) have argued that psychoanalysis and related schemes (for example, the theories of Jung and Adler) do not constitute properly formulated scientific theories (see also Kline, 1971). I shall therefore apply the criteria of theory construction of chapter 3 to a very brief evaluation of the claims of psychoanalysis and kindred doctrines. Psychoanalysts of various schools have (i) propounded hypotheses, and (ii) acted as applied scientists. The hypotheses adopted by many 'analysts' were those of Freud or close variants thereof (in the case of psychoanalysis), or those of Jung or Adler in other cases, or even eclectic systems of hypotheses. Freud postulated several major hypothetical constructs by asserting that the human personality comprises an 'ego', 'id', 'superego', 'censor', and so on (see Fenichel, 1945). Yet none of these hypothetical constructs is accompanied by precise and explicit statements expressing its properties in terms of a set of initial hypotheses of a theory.

We may be told that the 'id' provides the mature 'drive elements'. These theoretical constructs are allegedly 'unconscious', and hence not describable in language by the subject. The mature drive elements are postulated to be derived from infantile sexual 'drives'. What a 'drive' is supposed to be is itself very debatable, at least in Freudian (and similar) analytic theories. The male infant is postulated to have a normal 'unconscious' tendency to develop an *Oedipus complex* or 'mother-fixation', with associated unconscious hatred of his father. In addition there are postulated infantile 'unconscious' tendencies which in adult life become 'repressed'. The 'libido', another favourite Freudian hypothetical construct (lacking proper defining statements of initial hypotheses, but belonging probably to the 'id' class of hypothetical constructs), is supposed to be responsible for sexual pleasure-seeking, oral (and similar) tendencies, resulting sometimes in 'oral fixations' (or other appropriate fixations) which may lead to 'repressed complexes'. The drive elements of the 'id' are not only supposed to be unconscious but (by hypothesis) can sometimes be mutually opposed to each other, thereby producing an unconscious conflict.

If, in the course of this conflict one or the other of the 'drive elements' becomes repressed (or inhibited) by the 'superego' (which operates *via* the 'censor'), then these submerged elements tend, by hypothesis, to become diverted into abnormal 'unconscious channels', leading to overt signs of compulsion neurosis (such as neurotic tics, or 'urges to perform particular rituals'), or even to 'conversion phenomena' involving motor or other symptoms of hysteria. (*Cf.* Fenichel, 1945 for a detailed exposition of classical Freudian doctrine.)

It is not difficult to see that, because of their lack of explicit defining statements (initial hypotheses) from which deductions could be made, the Freudian hypothetical constructs ('ego', 'id', and so on) are metaphors. The 'unconscious', while not observable, does not function like an unobservable variable or like postulated structures of a properly defined theory (for example, in the sense in which DNA or messenger RNA functions in molecular biology, or like forces of classical dynamics). The 'unconscious' is as vague as the pronouncement of a mystic. (We know, for instance, that unconditioned reflexes occur by internal mechanisms of which we are not conscious. If we ascribed the occurrence of these reflexes to the operation of the 'unconscious', this would be as feasible and helpful as the use made of the 'unconscious' by psychoanalysts.) Anything an analyst wishes to explain can be assigned untestably and *ad hoc*, to the 'unconscious', which has therefore implicitly an unlimited number of *ad hoc* properties, which are ascribed as the need arises (*cf.* Bunge, 1967c, vol. 1, pp. 285–7, on the need for independent testability of the consequences of *ad hoc* hypotheses). It therefore provides a convenient reservoir for yielding unlimited *ignotum per ignotius* explanations.

The analyst claims that he can trace back neurotic conflicts (which he assumes to be present) to their unconscious source, by unmasking the unconscious repressed infantile drives, thereby allowing previously unconscious 'motivations' to become expressed in words. This, allegedly, exposes conflict-creating motivations to analyst and patient. According to Freudian theory the principal instruments for revealing unconscious conflicts are dreams and free associations. Freud's theory of dreams asserts that they are wish-fulfilments, and that dreams express these fulfilments *in symbolic form*, symbolising normally repressed unconscious motives (such as the Oedipus complex) which give rise to neurotic conflicts, but which in the dream find 'expression' in disguised form. By interpreting the dream symbols 'correctly' (whatever that may mean), the analyst is supposed to obtain the key for uncoding the unconscious conflict of the 'id' system, thus enabling him to resolve the conflict, or guiding the patient to do so.

As dream symbols do not appear as theoretical variables, it cannot be ascertained whether particular dream symbols represent in any sense deductions from any initial hypotheses of Freudian dream theory. In fact, dream analysis involves *ad hoc* interpretations of symbols, which could be interpreted in an indefinite number of alternative ways (as is, indeed, the case when different analysts, or different 'schools', such as Freudians and Jungians, analyse the same dream). The absence of testable empirical generalisations which are unambiguously deducible from (missing) Freudian initial hypotheses prevents indirect empirical anchorage of Freudian theory.

If a neurotic or hysterical patient shows signs of remissions of symptoms during or after a period of treatment, this is often accepted as evidence for the validity of the theory, including the correct interpretation of dreams or of 'free associations' by the analyst (which are *ad hoc*, since the theory does not allow us to deduce which interpretations are appropriate for any specific dream or any specific 'free associations'). If an analyst is asked how tranquilisers remove neurotic tics, he might say that they interfere with the 'id' system in a manner which we do not understand.

Spontaneous remissions of neurotic symptoms in the absence of treatment are known to occur, and failures of psychoanalytic treatment are frequent. As psychoanalysts have no adequate *statistical controls* (that is, of the progress of comparable treated *versus* untreated cases), it is not possible to rule out that many or most alleged cases of 'successful' analyses could be due to spontaneous remissions or other causes. Since this alternative interpretation has not been eliminated, the therapeutic efficacy of the psychoanalytical procedure remains open to grave doubts (*cf.* Eysenck, 1961; Bunge, 1967c). Again, why should different analysts (such as Jungians), who use quite different theories and interpretations, be able to claim comparable successes to the Freudians? There exists no compelling *independent empirical evidence* for the various theoretical assumptions of psychoanalysis apart from alleged therapeutic successes, and as the latter are open to doubt they cannot be chosen as validating criteria of the theory, particularly as the theory contains an unlimited number of inherent *ad hoc* hypotheses in dream interpretations.

Gellner (1959, p. 168) noted a striking similarity in outlook between the methodology of psychoanalysis and that of certain varieties of philosophical common-sense language analysis. He remarked with respect to the former:

> there, too, we have a doctrine and a technique in close symbiotic relationship. If we are interested in the technique and enquire about the statistics of its successes, we are hastily told that these in any case prove little: if we merely want a cure, why not go to Lourdes? The technique gives *insight* which is more important than therapeutic success alone. But the insight, one imagines, must have some connection with the doctrines and ideas. If one investigates the ideas which constitute the insight, and becomes worried by their vagueness, lack of confirmability and confirmation etc., one is assured—ah but the doctrine can only be understood in the light of the technique, the *practice*. So one shuttlecocks between the two . . . .

Although psychoanalytic theory has been claimed to be 'mechanistic', it fails to specify any underlying processes explicitly. Those who criticise the theory are said to be victims of some 'complex' which forces the criticism upon them as a defence against 'id' forces (*cf.* Ekstein, 1964, pp. 64ff, for further discussions which—somewhat unintentionally—serve to illustrate some of the preceding arguments).

4.3 General antitheoretical nature of metaphorical 'agent''

A. Metaphorical agents

Psychoanalysis belongs to a particular class of explanatory procedures which do not constitute properly constructed theories. These procedures rely on *metaphorical agents* (such as the 'ego', 'id', 'superego' or 'collective unconscious' of certain psychoanalytic schools). Metaphorical agents are hypothetical constructs lacking adequate accompanying initial hypotheses. Notably a vitalist will invoke metaphorical agents (see examples below), and pursue a dual strategy.

(i) He attempts to prove *a priori* that no conceivable 'mechanisms' could explain either embryogenesis or the mental processes of organisms. Various 'proofs' given by McDougall (1934), Driesch (1929) and others start from the assumption that specific types of mechanisms, namely those discussed by some particular vitalists, could not possibly account for morphogenesis or for the physical representations of mental events (e.g. Driesch, 1938; Bergson, 1911). It is tacitly taken for granted that the particular mechanisms considered by the vitalists are the only ones conceivable for all eternity (and hence that no mechanisms discovered in future could lead to explanations of the phenomena considered). After the vitalist has convinced himself of the soundness of his 'proof' he proceeds to his second objective.

(ii) He suggests 'explanations' couched in terms of metaphorical agents, which are explicitly claimed to be 'non-physical'.

Metaphorical agents are probably remnants of anthropomorphic ways of thinking—or 'demonology', as Needham (1942, p. 123) termed it. Needham remarked that

> For centuries science has struggled to rid itself of the remains of popular demonology. When Cambridge got its first Professor of Chemistry (in 1703), acids were male and alkalies female; minerals grew, like plants, from seeds; slaked lime protested by giving out heat; and solid bodies cleaved to themselves in cohesion because they preferred the touch of the tangible more than the feeble contact of air.

Typical metaphorical agents are McDougall's (1934) striving 'horme', Broad's (1925) 'psychic factor', H. H. Price's (1939) 'psychic aether', Driesch's (1929) 'entelechies', Kapp's (1951, 1955) 'diathetes', Eccles' (1953, chapter 8) 'influences', Teilhard de Chardin's (1960) 'omega point' and 'noosphere', and Bergson's (1911) '*élan vital*'. (Grant (1956) discussed Kapp's metaphysics in some detail, and Medawar (1961) gave an extensive critical evaluation of Teilhard de Chardin's (1960) book. Huxley (1942, p. 458) remarked that Bergson's '*élan vital*' is comparable to 'explaining' the mode of propulsion of a railway train in terms of an '*élan locomotive*'. The reader is also referred to Spence's (1944) remarks on hypothetical constructs which lack improper defining statements (initial hypotheses).)

The history of the metaphorical agent concept 'instinct' (see Beach's 1955 critical analysis) and the 'wholeness factor' postulated by some holists (*cf.* Allport, 1955, p. 113; Smuts, 1926; Goldstein, 1940) provide additional examples. (Madden's (1952) criticisms of particular formulations of *Gestalt* and holist 'theory', George's (1956, 1958) refutations of various apriorisms, Hebb's (1959) critical evaluations of (a) the interactionist hypothesis of Eccles (1953) and (b) the mentalism of J. Cohen (1958), Walshe and others, should be consulted. Lucid criticisms of vitalism were also provided by Lorenz (1950), Needham (1927, 1942, pp. 119ff), Hartmann (1933), Winterstein (1928), Schlick (1948) and Köhler (1938, p. 296), among others.)

Eccles (1953, p. 273), for example, postulated that his undefined 'influences' act upon cortical neurons, thereby increasing their probability for impulse discharges. Bergson asserted that the '*élan vital*' drove organismic development and evolution along directed pathways. Others postulated 'vital forces' which, when added to inanimate bodies, cause their behaviour and developmental processes, much as petrol, when added to an automobile, provides the driving couple *via* intervening machinery. 'Theorising' in terms of metaphorical agents appeals particularly to laymen, conveying to them the impression that they have received 'explanations' of complex phenomena, where in reality an *ignotum per ignotius* display has occurred (also well illustrated by Dunne's (1929, 1934) 'theories' of precognition).

Mentalism supplies other instances of metaphorical agents. Aristotle's 'psyche' and the 'soul' concept of religion (*cf. Brett's History of Psychology*, ed. Peters, 1962) are metaphors which may satisfy metaphysicians but play no part in scientific theories. Many religious metaphysicians argue that the 'soul' is responsible for man's actions, and attribute to it his moral responsibility. I suspect that this argument lies at the root of much opposition to mechanistic explanations of human cognition and behaviour. No doubt some metaphorical agents are very appropriate to religious sermons, but not to scientific discourse. As long as this is remembered, no clash can take place between science and religion. Confusion only occurs when vitalists, mentalists and others try to import religious (or other) metaphors into science.

Mentalists often transfer everyday modes of speech to new contexts where their common usage becomes meaningless. For instance, the old ideo-motor theory postulated that 'volition', 'striving' and 'ideas' could give rise to specifically patterned behaviour. The roots of mentalism lie in ordinary language, which uses some expressions *as if* they implied causal efficacy (see § 2.4B). The common usage of the word 'purpose' does not imply evidence for some hidden metaphorical agent which makes our muscles move so as to achieve specific goals. I need not labour these points, since Ryle (1949) has brilliantly dealt with similar topics.

B. Science, religion and brain-models

Science and religion differ fundamentally. While both are based on belief (because of the unprovability of hypotheses), religion assigns absolute reality status to its hypothetical metaphorical agents (God, or in some religions several gods, soul, holy ghost, and so on). These agents are not endowed with explicitly defined initial hypotheses from which empirically testable deductions can be made (for example, the statements 'God is all powerful', 'God is omniscient', and so on, are not intended to be tested by experiment). Scientific discourse is limited to providing empirically anchored explanations, while this does not apply to religious and everyday talk, which may command, evaluate, generate moods, moralise (see also Feigl, 1958, p. 417, for related arguments). Religion (and law) contain 'normative statements'. They assert what is 'right' or 'wrong', 'good' or 'evil' (and so on).

Scientific theories could explain how human brains structurally (that is, physically) represent values, but science *cannot decide* which values men ought to adopt. Likewise, brain-models (see p. 3) could explain how brains produce physical processes which represent evaluations in general, but no brain-model could decide which particular valuations made by living brains are *correct*. Brain-models can, for instance, only be expected to deduce from appropriately formulated initial hypotheses general procedures which brains could adopt to evaluate deductive systems. These models could therefore help philosophers of science, as well as neuropsychologists, to understand the *ability* of some human brains to assess the validity of *approximate* pathways which lead from initial hypotheses of particular theories to empirical generalisations. Yet brain-models *per se* could not 'justify' any particular assessment. Just as knowledge of the general principles of physiology and anatomy of an organism are necessary for the basic understanding of its possible abnormal functioning, while these principles *per se* are insufficient to 'justify' the diagnosis of any disease, so brain-models are necessary for understanding the cognitive mechanisms involved in producing scientific theorising, but *per se* cannot justify the procedures of theorising.

More generally, brain-models could only explain the range of cognitive and behavioural performances which nervous systems can control, and how this control could be effected. However, brain-models could never (without additional *ad hoc* hypotheses) justify that the use which brains make of their inbuilt mechanisms leads to results (decisions, hypothesis formulations, creative acts of all types, emotional states, evaluations, and so on) which are in any absolute sense the 'best possible'. In fact, the opposite is the case. Many human actions are inappropriate, plainly moronic or brutal (for example, Hitler's extermination of six million Jews).

As far as science is concerned, human rationality (as well as irrationality) can only be explained in general terms *via* brain-models. However, general brain mechanisms cannot justify the validity of any particular cognitive result obtained by them (such as any particular hypothesis or scientific procedure) any more than our understanding of the mode of action of a television system can justify the validity of a statement which is being transmitted in a television programme.

The preceding paragraphs, I believe, suggest that brain-models will enable us to understand the general biological machinery which is involved in our various scientific theorising activities. They should ultimately give us some idea how brains could accomplish deductive reasoning, acts of invention, like those leading to the development of inductive methodology. This is, I guess, as far as we can go in 'understanding' the foundations of science. Any pretence that the scientific enterprise *per se* is meaningful without human brains, and their general evaluatory and creative capacities, requires more 'justification' than the unjustifiable scientific inductions.

No scientist could either assert or deny that all events in the universe conform to mechanistic principles. Science can only postulate general classes of mechanisms, which (as a rule) can only be approximately applied to explain or predict the properties of particular systems in special cases. Generally science deals with class properties. To make statements about the mechanistic nature of *all* events in the world would require information about the initial conditions of the states of all its known (and its as yet unknown) systems and mechanisms at some particular locally prescribed times, and these conditions are in practice and in principle inaccessible. Hence the discourse of science is severely restricted. While this does not justify taking refuge in religion, there are no scientific grounds for opposing it.

The limitations of science mentioned also show why discussions of universal 'determinism' are metaphysical (that is, empirically undecidable). Similar remarks apply to such problems as individual 'freedom of action'. As we could never stipulate the initial states of all brain molecules (and the changes produced by food intake and interaction with the environment *via* sensory receptors, and so on), we could never define a *complete* system of interacting brain mechanisms even for a hypothetical brain. We can only explain major class properties of phenomena of psychology in terms of *postulated* autonomously operating classes of brain mechanisms. But this does not allow us to decide (a) that these mechanisms are either true, or (b) that they are the only ones which could operate.

Unlike Eccles (1953) I prefer, as far as possible, to explain psychological processes in terms of postulated autonomous brain mechanisms. However, I am not asserting that this implies (as it could not) complete mechanistic determination of thought processes (and so on). But I am stressing that issues of 'individual freedom' are empirically undecidable. This being so, we must for ever be left in doubt about the equally metaphysical issues of personal responsibility. A man could be a radical mechanistic scientist, and yet hold metaphysical (for example, religious) views on a wide range of empirically undecidable issues. Or he could be an agnostic. (See also § 1.8 for other relevant discussions.)

4.4 Organisms considered as self-organising automata

A. Automaton approach to biological systems

According to § 4.3B, we can only suggest general types of processes which organisms perform autonomously (subject to required precursor material, as in the case of protein synthesis). Hence we can say that, as far as we can ascertain at present, organisms behave in many respects like self-organising automata. Yet we could never conclusively either establish or disprove that they are *completely* self-organising automata.

Among those who try to explain organismic development and behaviour as far as possible in terms of autonomous and self-organising mechanisms, we find many different views, and even some surreptitious dogmatism has appeared. In the present state of biological sciences we must remain open-minded and try various approaches, even if some of us (e.g. Wassermann, 1972, 1973, 1974) preferentially explore particular types of mechanistic theories.

As reductionist approaches yielded rich results in the physical sciences, biologists and psychobiologists began to probe how far reductionism could lead in the life sciences. Earlier attempts to explain morphogenesis (Rashevsky, 1948; Turing, 1952) belong to the pre-molecular biology era. Turing's work was criticised by Waddington (1956), whose own theorising (Waddington, 1956, 1957, 1966) was confined to the formulation of nucleo-cytoplasmic feedback equations in single cells, and to cybernetic analogising (*cf.* Waddington, 1957—for example, his concept of 'epigenetic landscape'). Waddington (1957) could hardly be expected to resort to an extensive molecular biological control theory at a time when the relevant evidence was not available. Recently new trends of reductive theorising, about morphogenesis and related topics, have emerged. They suggest explanations of the genetic control of morphogenesis in terms of molecular biological theories. Some of these theories have only dealt with a few (Britten and Davidson, 1969; Davidson and Britten, 1971), others (Wassermann, 1972, 1973, 1974) with a wide range of phenomena. These trends suggest that reductive theories of biological phenomena are now becoming feasible in quite explicit and testable terms.

B. *Some notable misconceptions*

Earlier automaton approaches to biological or psychobiological phenomena (for example Rashevsky, 1948; Wiener, 1948, 1961; Craik, 1943; Sommerhoff, 1950) already favoured mechanistic interpretations (see also MacKay, 1952) but were confined to generalities or analogies. Numerous brain-models began to pursue this mechanistic trend. Their common aim is to explain major phenomena of experimental psychology and human everyday behaviour in terms of automaton theories. One group of theorists (e.g. Hebb, 1949; Wassermann, 1972, chapter 17, and 1974) believe that the psychobiology of brains should be explained in terms of known or conjectured structural and functional properties of neurons (and associated glial cells and their known substructures.)

Yet some mechanistically inclined behaviour theorists believe that a brain-model need only simulate input and output relations of brains or nervous systems, but not any properties of their components. As long as a non-brain-like automaton simulates a sufficiently wide range of typical human or animal behaviour, it provides in the opinion of theorists like Deutsch (1960), Fodor (1968, *cf.* p. 113–15) and others an *adequate* explanation of this behaviour. This view was contested by Gendron (1971) on the grounds that Deutsch as well as Fodor assume that (a) psychological (behavioural) constructs are *functional*, (b) neurological (anatomical) constructs are structural, and (c) functional constructs cannot be theoretically reduced to structural constructs. In fact, contrary to such opinions, neurological constructs, if properly formulated, involve structural *as well* as functional hypotheses (such as presumptive changes of states of specific macromolecular systems which encode memories; *cf.* Wassermann, 1974).

Let me cite Deutsch (1960) explicitly. He wrote (1960, p. 1):

> . . . there are two stages in a structural or neurophysiological explanation. The first stage is the devising of a system whose properties tally with our observations of behaviour. The second is the identification of the elements of this system in neural terms . . . . a psychological theorist need only be concerned with the first stage.

Again (Deutsch, 1960, p. 11) argued:

> . . . to suggest physiological mechanisms without direct observational warrant for their existence is fanciful. There is a great deal of substance in this objection but it cannot be treated as an objection to all kinds of attempts to arrive at a structural explanation. It applies only to a particular type of speculation—that which cannot in principle be checked by observations undertaken on the behavior of the animal as a whole or, in other words, the type of observations normally made by psychologists. These speculations concern the embodiment of the system employed. For instance, to attempt to guess at the particular change which occurs in the central nervous system during learning in the framework of a theory purporting to explain behavior is not only unnecessary but also purely speculative. That some type of change occurs may be inferred from the behavior of an animal. What this type of change is cannot be arrived at, nor is it very important for the psychologist to know . . . . to spectulate about terminal end boutons in the way that Hebb does or about changes of synaptic resistance seems to be trying to answer a question irrelevant, strictly speaking, to the psychological theorist. What behavior would one of these assumptions explain which the other would not?

In this passage Deutsch disregarded the possibility recognised by Mace (see § 1.5F) that there may exist many different levels at which we could explain natural phenomena. A neurophysiological explanation of behaviour (like a molecular explanation of the behaviour of gases) has goals differing from purely functional simulation (or 'explanation') of behaviour to which Deutsch aspired. Deutsch's type of simulation cannot achieve a theoretical linkage between neurobiology and experimental psychology—that is, it amounts to a proscription of theories which are more comprehensive than behaviour theory on its own. To call neurobiological theories 'speculative' simply means that they are not yet firmly empirically anchored and, like many other theories, have to be modified or replaced in order to augment their degree of empirical anchorage. Nobody would call classical dynamics 'speculative' because it does not apply to atomic physics.

Deutsch (1960, p. 12) continued the preceding passage as follows:

> This would seem a good argument against speculating about the mechanism underlying behavior, but not against attempting to infer the type of mechanism or the system producing behavior. Clearly, the question about what the actual physical change is which occurs during learning in the machine is the wrong type of question to ask and to attempt to answer . . . . Information about the physical identity of the parts of the machine sheds an extraordinarily feeble light on the explanation of the machine's capacities . . . .

This last passage is, I believe, on a par with Driesch's mistaken prophecies. It rests on the assumption that Deutsch can predict the deductive capacities of all future theories of brain function. People holding related opinions last century (and there were many) could have argued that it is wrong to ask questions about the internal systemic physical changes which take place in the photoelectric effect, or about the changes of states (of postulated electrons of atoms) of a gas which lead to the emission of specific spectral lines. However, from the behaviour of an excited hydrogen gas as a whole system, we could hardly be led to Schrödinger's equation. It was a combination of many data (including the Davisson and Germer (1927, 1928a, b) experiments—see Harnwell and Livingood (1933, p. 153)—of electron scattering by crystals or gratings) which led to confirmation of de Broglie's hypothesis and to the establishment of a consistent atomic physics, which explained molar phenomena in terms of unobservable quantum mechanical variables (such as Schrödinger functions).

One feels tempted to ask advocates of Deutsch's strategy how they could hope to establish a theory of genetic control of organismic development by ignoring molecular biology, and taking only the findings of molar sciences (such as genetics and non-molecular work in embryology) into account. This would correspond exactly to Deutsch's proposed procedure in a psychological theory which disregards brains. Deutsch (1960, p. 13) made his views quite explicit:

> The precise properties of the parts do not matter; it is only their general relationships to each other which give the machine as a whole its behavioral properties. These general relationships can be described in a highly abstract way . . . . Nevertheless, the machines thus made, will have the same behavioral properties, given the same sensory and motor side. Therefore, if we wish to explain the behavior of one of these machines, the relevant and enlightening information is about this abstract system and not about its particular embodiment. Further, given the system or abstract structure alone of the machine, we can deduce its properties and predict its behavior. On the other hand, the knowledge that the machine operates mechanically, electromechanically, or electronically does not help us very much at all.

To claim that we can deduce the *properties* of a machine or system is only relevant if the specific properties concerned cover a sufficiently wide range of known features. By listing only a narrowly restricted number of properties (namely those which are outwardly observable) one may leave many important problems unanswered. If, for example, a motor engineer wishes to repair an automobile or understand its working, familiarity with its internal structure (engine, gears, differential, and so on) becomes very relevant, and understanding of a turbine-driven automobile requires quite different knowledge from one driven by an internal combustion engine. (It is true that many drivers of automobiles regard their vehicles as black boxes, and are only familiar with their responses to manual manipulation.) Again, returning to the life sciences, in Sperry's experiments on split brains (see Gazzaniga, 1969, for a review) and in Lashley's (*cf.* 1950) experiments on memory storage systems in animals, the relation of behavioural responses to experimental neuroanatomical alterations was examined and led to very significant results. If we wish to understand the effects of actinomycin D on a circadian biological clock mechanism (*cf.* Bünning, 1964; Wassermann, 1972) it is not enough to use a simulator system (e.g. Pavlidis, 1971), but resort to molecular models is required (as actinomycin D is known to interfere with the genetic transcription mechanism, that is, with the synthesis of RNA macromolecules).

In line with Deutsch, some eminent workers used non-biological models to simulate biological behaviour (for example, Lorenz's (1950) hydromechanical model of 'innate behaviour', or Walter's (1953) 'tortoises' which exhibited 'choice behaviour' and conditioning). These gadgets were treated either as analogies (in the case of Lorenz) or to demonstrate that certain types of behaviour (which vitalists and others had claimed could in principle never be performed by mechanisms of any kind) could, in fact, be simulated by automata. Such models may help as first steps in clarifying our thinking about biological systems (see also Sommerhof, 1950). Yet gadget language is no substitute for explaining biological behaviour in terms of known (or conjectured) *biological* structures or ultrastructures (such as ribosomes or mRNA) with known or (conjectured) functions. (The word 'known' is used in the sense of theoretical knowledge.) Another serious weakness of Deutsch's type of gadget 'theorising' is its lack of extensibility (see § 3.5). The structural postulates of his theory cannot be combined with neurophysiological or biochemical postulates, since his structures are devoid of biological significance. His type of simulation is therefore *biologically* inadequate.

C. Different classes of automata

When scientists or philosophers debate whether organisms and their subcomponents (such as brains) are automata, confusion must be avoided by distinguishing different kinds of automata. First of all there are automata such as automobiles, television sets, aeroplanes (or Fernandéz-Morán's (1971) ingenious use of S. C. Collins' closed-cycle superfluid helium system which is coupled to an electron microscope) with *fixed numbers* of components. These machines can change the states of their components (a sparking plug, for instance, may be in a sparking or non-sparking state) but not the types or numbers of components. Electronic computers, for instance, are relatively inflexible, and their apparent versatility lies in the variety of man-made programmes which can be fed into them (although self-programming computers have been considered). Turing (1937a, b, 1950) established a general theory for certain classes of fixed component machines (see Gluskov, 1961, and Wang, 1963, for formal definitions of Turing machines), and electronic computers are special cases of Turing machines. Likewise, certain classes of postulated 'neural networks' (*cf.* Griffith, 1971) are Turing machines. A Turing machine operates on its input 'programme' by *systematic* procedures called *algorithms*. (Typical algorithms occur at a less sophisticated level in ordinary arithmetic, for example when we use a stepwise procedure in long division or in the multiplication of many-digit numbers. A more sophisticated molecular biological 'process algorithm' for a theory of organismic development was put forward by Wassermann,

1972, 1974.) The programme of a computer (or the tape which feeds into a Turing machine) provides detailed instructions for stepwise execution of the algorithm, or sequence of algorithms. Turing machines of certain kinds (such as computers) can often simulate the behavioural capacities of various fixed component automata.

I must now turn to another class of automata, namely *self-organising systems*. If submechanisms of organisms can be treated as automata (which is consistent with available evidence; *cf.* Wassermann, 1972), then such automata are of particular significance. They have the property that their component types and numbers are not fixed from the start, and that new component types can be self-generated by the automaton during its functioning or development. For instance, the development of higher organisms involves the formation of myriads of new cells (and new cell *types*—Sperry 1963; Wassermann, 1972, 1973, 1974), and the formation of memories by brains requires the building up of new engrams. Wound healing demands regeneration of missing cells and/or cell types. In a self-organising automaton some components, once established, could remain permanently fixed, while others undergo repeated modifications. The important point is that Turing machines are not self-organising automata, but some self-organising (and other) automata can in some respects function like Turing machines or be controlled by them. (Although the *output* of a computer may control industrial production of new parts of another machine, in doing so the computer does not change its own structure.)

While fixed component machines may respond differently to different environments (for example, different inputs) without essential changes (apart from wear and tear) of their structures, the internal structures (such as engrams) which self-organising systems build up may be environment-dependent. Thus, a typical mammal, when exposed to a specific antigen, will produce lymphocyte clone cells manufacturing specific antibodies directed against these antigens (*cf.* Burnet, 1959; Wassermann, 1972, chapter 16; Litwin *et al.*, 1973). The mammal, therefore, in response to the environmental antigenic stimulus, organises new cellular systems. (But the new cellular systems could partly arise from induced proliferation of already existing uniquely differentiated precursor cells; Wassermann, 1972, 1973, 1974). Again, bacteria may be induced to produce in large quantities protein types (inducible enzymes) which they normally only produce in trace amounts.

There exists no universally applicable automaton theory, because different branches of physics, chemistry or biochemistry are appropriate for dealing with different components of different automata. It is therefore not very profitable for theoretical biologists to argue about mechanisms in general terms in the way that some cyberneticians enjoy doing. To say that a certain type of behaviour (for example, human behaviour) is produced by negative feedback is not illuminating unless the precise biological or biochemical mechanisms involved and their precise modes of operation are postulated. Otherwise, general feedback diagrams function as obscure black boxes. Similar remarks apply to the favourite notions of 'integration' and 'complexity of neuron connections' which are bandied about by some psychologists and neurologists.

D. Thought processes and automata

Many machines are not designed to think (George, 1956, p. 244), and it is now recognised that electronic computers or Turing machines *per se* do not think, but operate on programmes. Creative thinking involves planning, that is, the construction of meaningful complex *representations* by brains of such things as musical compositions, mathematical theorems, novels, plays and technological inventions. Brains can use algorithms (as shown by long division, and so on), but much brain activity is nonsystematic and *heuristic*; that is, it is not based on algorithmic procedures *only* (*cf.* Wertheimer, 1959). Heuristic behaviour is characteristic of self-organising rather than fixed component automata. Although a Turing machine can operate on a heuristic programme, this does not imply that the machine itself operates heuristically. The heuristic programme supplied to the computer is generated by a human brain (at least in most contemporary uses of computers).

(Hesse (1968), while reviewing Scheffler's (1967) book, referred to the 'paradox of meaning', according to which no term *P* in a theory T_1 can be said to mean the same in any other theory T_2 (for example, the term 'energy'). On the basis of this it was argued that *logically* theories cannot be corrected or improved, but that there can be a sequence of theories which are unrelated by any logical or rational ties. In reaching this conclusion Hesse (in common with others) disregarded the point made earlier that theories are only symbols on paper (as are novels or symphonies) until interpreted and acted upon by human brains. Human brains, though able to operate on the symbolism of mathematical logic, have also capacities which go far beyond mathematical logic in constructive thinking, by employing heuristics. Hence, while different theories (constructed by brains) may use the same names for things which are differently defined in each theory, the similar sounding concepts, *while not logically related*, *could be heuristically connected by brains*. Brains could discover similarities of functional concepts (such as energy) even in different theories. The energy concept in Einstein's relativity theory is not essentially dissimilar to the energy concept of Newtonian mechanics or of quantum mechanics, although formally (that is, in terms of postulatory structures) the theories may differ drastically. The 'paradox of meaning' disappears when it is appreciated that brains, unlike formal theories on paper, have the capacities to establish class similarities. In fact, without brains not even formal theories would make sense. The paradox arises from the faulty notion that we must only admit for certain purposes algorithmic (and logical) capacities of brains, but not their abilities to detect similarities or to compare and connect old with new ideas.)

Our assessment of a presumptive automaton (brain, and so on) as a thinking system depends on arbitrary criteria which must be accepted by others. I suspect that the most plausible way of achieving this is by developing biochemically and biologically adequate brain-models which, according to their theoretical specification (the models being paper-and-pencil theories), can perform tasks ranked equal to the most complex intellectual achievements of man. In this context it is again useful to adhere to Minsky's dictum (see p. 7). When asked what we mean by thinking, we chose our answer in terms of sufficiently complexly designed brain-models (that is, theories). As a decision procedure for the adequacy of these theories (relating brain function and cognitive processes) we can invoke the judgment of fellow scientists to see whether they agree that the capacity of the brain-model approaches that of sophisticated men. (This 'criterion of appeal to the scientific community' is not new and has been suggested by others when questions were raised whether machines can 'think'.)

We must remember that the range and complexity of human thought processes may vary widely for different individuals. A man with an IQ of 90 can form grammatically and serially ordered, properly conceptualised (that is, meaningful) sentences. This feat, while far exceeding the capacities of an ape, is trivial compared to the subtleties of man's most sublime creations. Hence, when we consider 'human thinking', comparisons based on 'our own language rules', namely those used (competently) by the 'man in the street' (for example, as proposed by Midgley, 1955) are unprofitable. Comparisons are only useful if they include the most sophisticated performances of man (such as abstract mathematical arguments, subtle forms of musical composition, and so on).

It is often believed that all thinking is conceptual. But this seems very improbable. The fact that *language* utilises concepts as a means of symbolising objects, events, moods, values, and so on, does not mean that brains represent these things only in conceptual form. Evidence based on perception (for example, visual figural aftereffects—see Spitz, 1958 for a review, and Hochberg, 1971—eidetic images, visual illusions, and many other 'field phenomena') suggests that brains can provide detailed transformed representations of retinally (and so on) represented 'images'. (The 'transformations' must produce 'depth', for example from monocular 'cues', such as Gibson's (1950) texture density gradients, and account for perceptual constancies and other properties of brain representations.) Thinking, like perception, occurs at the concrete and conceptual level. The genesis of explicit (i.e. 'concrete') representations, concept formation, the establishment of new relations between concepts, the establishment of new engrams, this and much more is required in constructive thinking. In fact, when we are planning to reach a particular destination, and plan our route, we must have a detailed (and not just a conceptual) memory of the route and its landmarks. A representational painter (as distinct from an abstract one) puts on canvas not concepts but ordered and precise configurations which are meant to correspond to those actually perceived (and actually do so remarkably well).

Some antimechanists have equated the word 'automaton' with a Turing machine, thereby disregarding the difference between self-organising systems (such as brains) and fixed component machines (e.g. Mays, 1952; see the reply to his paper by George, 1956). Lucas (1961) thought that Gödel's (1930) theorem rules out that mental processes can be machine-like. (Gödel's theorem states that, in any *consistent* formal system which makes it possible by use of an algorithm to deduce (or 'prove') specific classes of formulae from given axioms, there will exist formulae of that class which cannot be proved by means of that algorithm.[1]) Lucas supposes that a formula *F* is not provable in a system, but that formula *F* must be meaningful to a 'rational being', namely man, since that being could otherwise not decide that the formula is unprovable in the system. From this Lucas concluded that 'minds' or 'rational beings' cannot be machine-like, as they can understand and make a decision about the status of a formula which a Turing machine could not make. I shall only add a few arguments to the discussion between Lucas and his critics (Good, 1968; Whitley, 1962; Benacerraf, 1967; Lucas, 1968a, b; Hanson, 1971).

(i) Lucas disregarded that there exist also self-organising automata, which are not Turing machines, and to which Gödel's theorem does not necessarily apply, or if it applies its form ought to be stated (*cf.* George, 1962).

(ii) As it is impossible to state initial states for brains because these systems arise developmentally (*cf.* Wassermann, 1972) we cannot decide by any criteria whether brains form automata which conform to Gödel's theorem. Lucas' argument is therefore concerned with a metaphysical (that is, an undecidable) issue. What brains could achieve can only be decided by exploring appropriate brain-models, and not by metaphysical apriorisms based on arguments which disregard developmental and structural features of actual brains. Scientists are concerned with actual situations and not with retorts to unanswerable metaphysical questions (posed, for example, by Lucas, 1968b).

[1] Lucas' (1961) presentation of Gödel's theorem can be summarised as follows. We consider the formula *F* which says 'this formula is unprovable in the system'. If *F* were provable in the system the formula *F* could not be applied to itself without contradiction, so that formula *F* would either not be provable in the system or be false. Moreover, if *F* were provable in the system it would have to be true, since in any consistent system nothing false can be deduced from the axioms. Hence if, on the assumption that *F* were provable in the system, *F* would turn out to be false, whereas *F* could not be false if the system is consistent, we conclude that *F* is unprovable in the system.

4.5 Probabilistic behaviour and probability computers

It is often mistakenly assumed that an automaton capable of exhibiting choice behaviour, and which at the molar level is ranked as a 'stochastic system', must be a *probability computer*. This fallacy is deeply rooted among some contemporary thinkers and has had important consequences. As man and animals exhibit choice behaviour, and for other reasons, some brain-model theorists (such as Uttley and Rosenblatt, and others) have concluded unjustifiably that brains operate as probability computers. However, it is not difficult to see that a system which *conforms* to probability laws *does not have to compute probabilities*. Atoms, molecules and elementary particles (such as electrons) which conform to the hypotheses of quantum mechanics *satisfy* quantum mechanical probability distributions which can be calculated by man. But the atoms (and so on) themselves do not compute probabilities. Brains are composed of macromolecules and smaller molecules, ions, and so on, which may *behave* probabilistically without being probability computers, and molar choice behaviour of man and animals could be (but I suspect is not!) a consequence of the probabilistic behaviour of constituent brain molecules, so that we do not have to assume that brains operate as probability *computers*.

4.6 Brains as pattern recognisers and elimination of physicalism

Because of the prevalence of people whose only scientific education occurs in schools, it was (and is) widely believed that pointer readings (or related scale coincidences) form the essence of scientific observation (Schlick's 1935 essay, for example, defended this view). People often do not realise that the crux of public observation consists of *comparisons* of observed things with standards. Pattern recognisers could achieve such comparisons by matching a pattern against a standard configuration. (In this sense, enzymes are pattern recognisers of substrate molecules.) Pattern recognition could involve deformable templates. Brains, when operating as pattern detectors, can recognise and classify gay or sad faces (by comparing them with engrams), identify objects and perform vastly more abstract types of pattern recognition (such as the recognition of a movement of a symphony). (See also pp. 64 and 68.)

'Physicalists' who, in the tradition of J. B. Watson (1919), did not pay attention to the role of pattern recognising capacities in the detection of physical properties (*cf.* Neurath (1931); Carnap's (1935, 1938a) earlier writings; and Bergmann's (1956) remarks on physicalism) thought mistakenly that we could not talk of happy or sad faces, as these were not 'physical' properties, and that we had to speak instead of measured distributions of light intensities (for example, when a photo of a happy face is being examined). Such arguments disregarded the fact that human brains are not only used as pattern recognisers but also as evaluating systems. Physicalism was based on the already criticised view (§ 1.3) that human records (printed, written, photographed, tape recorded, and so on) are meaningful *per se* without interpretative systems, namely brains. The physicalists avoided invocations of brains, because they thought it best to leave the (then) largely unknown alone, and to bypass something which cannot really be bypassed. They went over the cliff in order to avoid an obstacle on the road.

4.7 Brain-models and subjective experiences

A. General arguments

Some philosophers and some antimechanists have stressed that brain-models and related mechanistic studies are unable to *explain* private experiences (*cf.* Polanyi, 1961; Hirst, 1959, pp. 210–11). True enough, as brains are physical systems, they could not be expected to provide anything but the physical basis of private experiences. The fact that private experiences are publicly unobservable would by itself be no obstacle to a theoretician, since earlier discussions in this book showed that scientific theories contain many unobservables (such as electron spin, 'strangeness' parameters of elementary particles; *cf.* Schweber, 1964, pp. 286–7, although physicists refer to these *technically* often as 'observables'). These unobservables can nevertheless explain indirectly a number of observable properties of systems. People might be tempted to argue that possibly with each elementary particle there are associated certain specific unobservables (variables), whose changes of state-values give (epiphenomenally) rise to private experiences. On this view consciousness would not be an emergent property but could, for all we know, be a property which accompanies specific changes of states of some or all elementary particles.

However, such arguments would not get us very far, except for suggesting that the point of view according to which conscious experiences *per se* are inseparably linked as emergents to *complex* forms of material organisation might not be justified. But more cannot be said. The argument is bound to remain metaphysical, since only systems as complex as man seem capable of communicating in symbolic terms that they are conscious. Hence, even if elementary particles could reach states with conscious epiphenomena, they would be unable to communicate this in a form meaningful to man.

Accordingly, epiphenomenalism even in its more complex form must remain a metaphysical hypothesis. We have no reasons for denying that certain physical processes could be *accompanied* (but not caused) by private experiences (*cf.* George, 1956, p. 246), but we cannot decide the validity of epiphenomenalism (see below). Even if epiphenomenalism is accepted, it must be understood that brain-model theorists are concerned only with explaining *physical* processes (some of which may epiphenomenally be accompanied by private experiences). We cannot reproach them for failing to explain private experiences *per se*, as scientific explanations of metaphysical issues are not possible.

B. *Some remarks on interactionism*

Some 'mentalist' philosophers (e.g. Scriven, 1966) have tried to 'prove' that conscious mental states are *causes* of behaviour. Scriven's (1966) 'proof' runs as follows:

> Having a serious toothache (a 'conscious state') causes me to go to the dentist . . . The problem is to show the nonredundancy of the pain-feeling, i.e., to show that the remaining set of conditions will not of themselves produce the dentist-going behavior. *If the remaining conditions had to include the concurrent brain-state, this would raise an interesting difficulty. They do not* and hence interactionism is proved. (Italics are mine.)

Scriven's conclusion hinges on the 'interesting difficulty' which would arise if the 'remaining conditions' would include concurrent brain-states. His statement that concurrent brain-states are not included among the remaining conditions *is an assumption but no proof*, and requires justification if his statement is to be regarded as a 'proof'. He states that

> The interesting difficulty is this: our assumptions [of a one–many or one–one mind–body correlation, and a complete body-level explanation of overt behavior] assure us of a brain correlate of pain, and that the brain-story is enough to explain, in a respectable scientific way, the trip to the dentist. This appears to show that we could dispense with the pain, and still get the same overt behavior. This would show that the pain is causally inefficacious, and hence that epiphenomenalism and not interactionism is true.

Up to this stage Scriven admits that purely autonomously operating brain mechanisms could suffice for explaining behaviour in a manner consistent with epiphenomenalism. But at this stage he refers to an argument by Baier (1962), and adapts it to add to the list of apriorist 'proofs' (see p. 130). He carefully chooses his assumptions, apparently in the tacit belief that they are *a priori* justified, and that no equally plausible counterhypotheses are possible.

Scriven argued:

> Suppose we discover a drug or an operation which *in fact* cut out the pain but still left a connection such that states of advanced decay in teeth led to our going to the dentist. Wouldn't this show that the pain is an epiphenomenon? Not at all; it would only show that *another* brain-state besides the one correlated with the pain can initiate dentist trips, i.e., it demonstrates plurality of causes, not the inefficacy of an alleged cause. The . . . assumption [of a one–many or one–one mind–body relation] guarantees that we can't have the same brain-state both with and without the pain, so a fortiori we can never get the pain-associated brain-state, without the pain, leading to or explaining the dentist-visiting outcome.

Scriven's argument is based on a 'thought experiment' which could not conceivably yield an unambiguous answer favouring interactionism. Granted that activations of pain centres are obviously not the only possible causes for prompting visits to dentists (for example, in the case of visits for a routine check-up), the interactionist would have to provide *unambiguous* evidence (without possible experimental artefact) that, with inactivation of, or in the absence of, those pain centres which normally respond to dental nerves, the dental visits are definitely caused by caries acting on these dental nerves. As Scriven admits the possibility of multiple *central* nervous causation of dental visits, a dental visit in the presence of caries (but with inactive or absent dental pain centres) could be due to a central activity (such as a tendency to have a routine check-up) which has nothing to do with the caries acting on the dental nerve. I conclude (as we could have expected) that metaphysical doctrines like interactionism and epiphenomenalism, cannot be empirically confirmed or refuted.

C. Epilogue: brains and physical fields

Neuropsychologists are often asked: what assures that particular cell types (that is, specific neurons) are under suitable conditions associated with conscious states, while there is no apparent evidence that other cell systems are directly involved in conscious experiences? There could be many conceivable answers. If neurons are each unique in the sense that each neuron synthesises a set of cell-unique protein types under strictly genetic control (Wassermann, 1972, 1973, 1974), then only certain neurons which synthesise appropriate cell-unique protein types might be associated with conscious experiences.

It is also possible that the 'true carriers' of conscious experiences are not the neurons *per se* but particular (yet to be discovered) elementary particles which (in second quantisation) are associated with physical fields (conceivably gravitons of rest mass zero and spin 2?) which only become bound to particular proteins of specific neurons. If this purely speculative conjecture were correct, then these field-associated particles (and hence the fields which carry them) might become selectively activated through prior activation of the specific proteins to which they are bound, and the ensuing conscious state *could be an epiphenomenon of the field activation*. I am not suggesting that we should take this type of speculation seriously at the moment. However, it shows that the hypothesis that brain constituents *per se* must be the physical representatives of conscious experiences could be an inappropriate hypothesis, and that the brain molecules might only act as necessary intermediate coupling systems with the constituents of certain physical fields which (when activated) are the true *physical representatives* of subjective experiences. Epiphenomenalism is compatible with the latter alternative (but see p. 144). (Addendum: With respect to the reinstatement of 'knocked-out' theories, an excellent example is provided by Pecker *et al.*, 1972; see also Anon, 1972, *Nature*, **237**, 183.)

APPENDIX: SOME ESSENTIALS OF VECTOR ANALYSIS

I shall confine the following note on vector analysis to a few essentials, and shall not pretend to provide a complete axiomatic treatment, as many important definitions and formulae will not be given.

A *localised* 'uninterpreted' vector

$$\boldsymbol{a} = (a_1, a_2, a_3) \tag{A.1}$$

is defined by a triplet of real numbers (a_1, a_2, a_3) which may be called the 'coordinates' of a point in a right-handed rectangular coordinate system with origin defined by the zero vector (0, 0, 0). In this coordinate system the 'coordinates' are given only by numbers which are not associated with any 'units' (such as units of length). This 'unit-free' system differs from familiar Cartesian systems which are applied to Euclidean geometry, and for which the three numbers (a_1, a_2, a_3) are associated with units of length. The magnitude of the vector (A.1) is defined by

$$|\boldsymbol{a}| = (a_1^2 + a_2^2 + a_3^2)^{\frac{1}{2}} \tag{A.2}$$

The vectors

$$\boldsymbol{i} = (1, 0, 0), \qquad \boldsymbol{j} = (0, 1, 0), \qquad \boldsymbol{k} = (0, 0, 1) \tag{A.3}$$

are called 'unit vectors', since their magnitudes are equal to one. Multiplication of a vector $\boldsymbol{a}$ by a multiplier p is defined by

$$p\boldsymbol{a} = (pa_1, pa_2, pa_3) \tag{A.4}$$

In (A.4) p is a real number which may *either* be regarded simply as uninterpreted *or* be given an interpretation by being associated with a 'unit' of some physical 'dimension'. Addition of two (uninterpreted) vectors $\boldsymbol{a} = (a_1, a_2, a_3)$ and $\boldsymbol{b} = (b_1, b_2, b_3)$ is defined by

$$\boldsymbol{a} + \boldsymbol{b} = (a_1 + b_1, a_2 + b_2, a_3 + b_3) \tag{A.5}$$

and subtraction is defined by replacing the plus signs by minus signs in (A.5).

From (A.3) and (A.4) it follows that

$$a_1\boldsymbol{i} = (a_1, 0, 0) \tag{A.6}$$

Similarly, it follows from (A.3), (A.4) and (A.5) that

$$a_1\boldsymbol{i} + a_2\boldsymbol{j} + a_3\boldsymbol{k} = (a_1, 0, 0) + (0, a_2, 0) + (0, 0, a_3) = (a_1, a_2, a_3) = \boldsymbol{a} \tag{A.7}$$

showing that any vector can be expressed as a suitably multiplied superposition of the three unit vectors.

(A.4) has another important implication. If we wish to *interpret* a vector as some physical entity, then we can choose the multiplier p in (A.4) to be of numerical magnitude 1 but to have an appropriate 'dimension' (that is, p could be a unit magnitude of length, or a unit magnitude of momentum, or a unit magnitude of force, and so on; see p. 87 above). Vector addition can only be performed if the vectors in (A.5) are either both uninterpreted or if they are both multiplied by a 'unit quantity p' of the same dimension. In other words, a vector of a given physical dimension can only be added to another vector of the same physical dimension. The magnitude of an interpreted vector $\boldsymbol{a}$ will, according to (A.2), be $p\,|\boldsymbol{a}|$ where p is numerically equal to 1, but has an appropriate 'dimension'.

The vector product between two vectors $\boldsymbol{a}$ and $\boldsymbol{b}$ is defined by

$$\boldsymbol{a} \wedge \boldsymbol{b} = (a_2b_3 - a_3b_2, a_3b_1 - a_1b_3, a_1b_2 - a_2b_1) \qquad \text{(A.8)}$$

where the vectors $\boldsymbol{a}$ and $\boldsymbol{b}$ could have different 'dimensions' (that is, different physical interpretations). The '*distance*' d between two points Q and Q' whose positions in the same vector space are given by the vectors $\boldsymbol{a} = (a_1, a_2, a_3)$ and $\boldsymbol{a}' = (a'_1, a'_2, a'_3)$ respectively is defined (see (A.2)) by

$$d = |\boldsymbol{a} - \boldsymbol{a}'| = [(a_1 - a'_1)^2 + (a_2 - a'_2)^2 + (a_3 - a'_3)^2]^{\frac{1}{2}} \qquad \text{(A.9)}$$

where (A.5) has been used to define the difference between the two vectors. By interpreting the vectors in terms of physical 'dimensions', it is seen that the '*distance*' defined by (A.9) need not correspond to spatial distance, but has the same 'dimensions' as the vectors. Thus if $\boldsymbol{a}$ and $\boldsymbol{a}'$ in (A.9) denote spatial position vectors (measured in length), then d has the dimension of a length, but if $\boldsymbol{a}$ and $\boldsymbol{a}'$ have dimensions of forces, then the 'distance' will have the dimension of a force.

For a scalar field function $\varphi(x_1, x_2, x_3, t)$, where x_1, x_2, x_3 are spatial coordinates (having the dimensions of lengths) and t denotes the time, we define the gradient of φ by

$$\text{grad}\,\varphi = \nabla\varphi = (\partial\varphi/\partial x_1)\boldsymbol{i} + (\partial\varphi/\partial x_2)\boldsymbol{j} + (\partial\varphi/\partial x_3)\boldsymbol{k} \qquad \text{(A.10)}$$

where $\boldsymbol{i}, \boldsymbol{j}$ and $\boldsymbol{k}$ are given by (A.3).

For a vector field $\boldsymbol{A} = (A_1, A_2, A_3)$, where A_1, A_2 and A_3 are each functions of x_1, x_2, x_3 and t, we define the vector curl $\boldsymbol{A}$ by means of

$$\text{curl}\,\boldsymbol{A} = [(\partial A_3/\partial x_2) - (\partial A_2/\partial x_3)]\boldsymbol{i} + [(\partial A_1/\partial x_3) - (\partial A_3/\partial x_1)]\boldsymbol{j} + [(\partial A_2/\partial x_1) - (\partial A_1/\partial x_2)]\boldsymbol{k} \qquad \text{(A.11)}$$

BIBLIOGRAPHY

Achinstein, P., 1968, *Brit. J. Phil. Sci.*, **19,** 159.
Ackermann, R., 1961, *Phil. Sci.*, **28,** 152.
Allport, F. H., 1955, *Theories of Perception and the Concept of Structure*, New York: Wiley.
Allport, G. W., 1954, *The Nature of Prejudice*, Reading, Mass.: Addison-Wesley.
Alvarez, W., 1972, *Nature Physical Science*, **235,** 103.
Arbib, M., 1964, *Brains, Machines and Mathematics*, New York: McGraw-Hill.
Ayer, A. J., 1956, *The Problem of Knowledge*, Harmondsworth: Penguin.

Baier, K., 1962, *Australasian J. Phil.*, **40,** 1.
Bartley, W. W., 1964, in *The Critical Approach to Science and Philosophy* (ed. M. Bunge), p. 3, London: Collier-Macmillan.
Beach, F. A., 1955, *Psychol. Rev.*, **62,** 401.
Becker, P. E. (ed.), 1964, *Humangenetik*, vols. 1–5, Stuttgart: Thieme.
Benacerraf, P., 1967, *Monist*, **51,** 9.
Berenda, C. W., 1950, *Phil. Sci.*, **17,** 123.
Bergmann, G., 1956, *Psychol. Rev.*, **63,** 265.
Bergson, H., 1911, *Matter and Memory* (trans. from the French by N. M. Paul and W. Scott Palmer), London: George Allen & Unwin.
Bernays, P., 1964, in *The Critical Approach to Science and Philosophy* (ed. M. Bunge), p. 32, London: Collier-Macmillan.
Blake, C. C. F., Evans, R. R., and Scope, R. K., 1972, *Nature New Biology*, **235,** 195.
Bohm, D., 1964, in *The Critical Approach to Science and Philosophy* (ed. M. Bunge), p. 212, London: Collier-Macmillan.
Born, M., 1927, *The Mechanics of the Atom*, London: Bell.
Born, M., 1934, *Proc. Roy. Soc. A*, **143,** 410.
Born, M., 1948, *Atomic Physics* (4th edition), London: Blackie.
Born, M., 1949, *Natural Philosophy of Cause and Chance*, Oxford: Clarendon Press.
Born, M., 1961, in *Werner Heisenberg und die Physik unserer Zeit* (ed. F. Bopp), Braunschweig: F. Vieweg & Sohn.
Born, M., and Green, H. S., 1949, *A General Kinetic Theory of Liquids*, Cambridge: Cambridge Univ. Press.
Born, M., and Infeld, L., 1934a, *Proc. Roy. Soc. A*, **144,** 425.
Born, M., and Infeld, L., 1934b, *Proc. Roy. Soc. A*, **147,** 522.
Born, M., and Infeld, L., 1935, *Proc. Roy. Soc. A*, **150,** 141.
Born, M., and Jordan, P., 1930, *Elementate Quantenmechanik*, Berlin: Springer.
Born, M., and Wolf, E., 1959, *Principles of Optics*, London: Pergamon.
Brain, W. R., 1941, *Brain*, **64,** 43.
Brain, W. R., 1951, *Mind, Perception and Science*, Oxford: Blackwell.
Braithwaite, R. B., 1953, *Scientific Explanation*, Cambridge: Cambridge Univ. Press.
Bridgman, P. W., 1927, *The Logic of Modern Physics*, New York: Macmillan.
Britten, R. J., and Davidson, E. H., 1969, *Science*, **165,** 349.
Britton, K., 1953, *John Stuart Mill*, Harmondsworth: Penguin.
Broad, C. D., 1925, *The Mind and its Place in Nature*, London: Kegan Paul, Trench, Trubner and Co.
Brody, B. A., 1970, *Readings in the Philosophy of Science*, Englewood Cliffs, N.J.: Prentice-Hall.
Brown, G. B., 1952, *Science: its Method and its Philosophy*, London: Allen & Unwin.
Brown, G. B., 1956, *Sci. Prog.* (London), **44,** 619.
Brown, G. S., 1957, *Probability and Scientific Inference*, London: Longmans, Green & Co.
Brown, W., 1936, in *Proc. 25th Anniv. Celebr. Inaug. Grad. Stud.*, p. 116, Los Angeles: Univ. of South California Press.

Buck, R. C., and Cohen, R. S. (eds.), 1971, in *Boston Studies in the Philosophy of Science*, vol. 8, Dordrecht: D. Reidel Publ. Co.
Bunge, M., 1959, *Causality: The Place of the Causal Principle in Modern Science*, Cambridge, Mass.: Harvard Univ. Press.
Bunge, M., 1964, in *The Critical Approach to Science and Philosophy* (ed. M. Bunge), p. 234, London: Collier-Macmillan.
Bunge, M., 1967a, in *Studies in the Foundations, Methodology and Philosophy of Science* (ed. M. Bunge), vol. 2, pp. 1–6 and 105, Berlin: Springer.
Bunge, M., 1967b, *Foundations of Physics*, Berlin: Springer.
Bunge, M., 1967c, in *Studies in the Foundations, Methodology and Philosophy of Science* (ed. M. Bunge), vol. 3, parts 1 and 2, Berlin: Springer.
Bünning, E., 1964, *The Physiological Clock*, Berlin: Springer.
Burnet, F. M., 1959, *The Clonal Selection Theory of Acquired Immunity*, Cambridge: Cambridge Univ. Press.
Burt, C., 1958, *Brit. J. Psychol. Statist. Sect.*, **11,** 31.
Burt, C., 1958a, *Brit. J. Psychol. Statist. Sect.*, **11,** 77.
Burt, C., 1959, *Brit. J. Psychol. Statist. Sect.*, **12,** 153.
Butchvarov, P., 1968, *Phil. Sci.*, **35,** 292.
Byerly, H., 1969, *Brit. J. Phil. Sci.*, **20,** 135.

Čapek, M., 1969, in *Boston Studies in the Philosophy of Science* (ed. R. Cohen and M. W. Wartofsky), vol. 5, p. 400, Dordrecht: D. Reidel Publ. Co.
Carnap, R., 1929, *Abriss der Logistik, mit Besonderer Berücksichtigung der Relationstheorie und ihrer Anwendung*, Wien.
Carnap, R., 1935, *Philosophy and Logical Syntax*, London: Kegan Paul.
Carnap, R., 1938a, in *International Encyclopaedia of Unified Science*, **1,** no. 1, Chicago: Chicago Univ. Press.
Carnap, R., 1938b, in *International Encyclopaedia of Unified Science*, **1,** no. 3, Chicago: Chicago Univ. Press.
Carnap, R., 1946, *Science*, **104,** 520.
Carnap, R., 1951, *Logical Foundations of Probability*, London: Routledge and Kegan Paul.
Carnap, R., 1966, in *Mind, Matter and Method* (ed. P. K. Feyerabend and G. Maxwell), p. 248, Minneapolis: Univ. of Minnesota Press.
Carnap, R., 1956, *Meaning and Necessity: A Study in Semantics and Modal Logic*, revised edition (Phoenix Books), reprinted 1964, Chicago: Chicago Univ. Press.
Cattell, R. B., 1950, *Personality*, New York: McGraw-Hill.
Clarke, M. L., 1958, *Brit. J. Psychol. Statist. Sect.*, **11,** 75.
Cleave, J. P., 1970, *Brit. J. Phil. Sci.*, **21,** 269.
Cohen, J., 1958, *Humanistic Psychology*, London: Allen & Unwin.
Colonnier, M., 1968, *Brain Res.*, **9,** 268.
Condon, E. U., and Shortley, G. H., 1935, *Theory of Atomic Spectra*, Cambridge: Cambridge Univ. Press.
Cotton, J. W., 1955, *Psychol. Rev.*, **62,** 303.
Coulson, C. A., 1961, *Valence* (2nd edition), Oxford: Oxford Univ. Press.
Craik, K. J. W., 1943, *The Nature of Explanation*, Cambridge: Cambridge Univ. Press.
Crick, F. H. C., Barnett, L., Brenner, S., and Watts-Tobin, R. J., 1961, *Nature*, **192,** 1227.

Davidson, E. H., and Britten, R. J., 1971, *J. Theoret. Biol.*, **32,** 123.
Davisson, C. J., and Germer, L. H., 1927, *Phys. Rev.*, **30,** 705.
Davisson, C. J., and Germer, L. H., 1927a, *Proc. Natl. Acad. Sci. U.S.*, **14,** 317.
Davisson, C. J., and Germer, L. H., 1927b, *Proc. Natl. Acad. Sci. U.S.*, **14,** 619.
De Busk, A. G., 1968, *Molecular Genetics*, New York: Macmillan.
Deutsch, J. A., 1960, *The Structural Basis of Behaviour*, Cambridge: Cambridge Univ. Press.
Dicke, R. H., and Wittke, J. P., 1963, *Introduction to Quantum Mechanics*. Reading, Mass.: Addison-Wesley.
Dirac, P. A. M., 1958, *The Principles of Quantum Mechanics* (4th edition), Oxford: Clarendon Press.

Driesch, H., 1929, *The Science and Philosophy of the Organism* (2nd edition), London: Black.
Driesch, H., 1938, *Alltagsrätsel des Seelenlebens*, Stuttgart: Deutsche Verlagsanstalt.
Duhem, P., 1954, *The Aim and Structure of Physical Theory* (1914) (trans. from the French by P. P. Wiener), Princeton N.J., Princeton Univ. Press (republished 1961, New York: Atheneum Press).
Dunne, J. W., 1929, *An Experiment with Time*, London: Black.
Dunne, J. W., 1934, *The Serial Universe*, London: Faber and Faber.
Du Praw, E. J., 1965, *Nature*, **206**, 338.

Eccles, J. C., 1953, *The Neurophysiological Basis of Mind*, Oxford: Clarendon Press.
Eccles, J. C., 1964, in *The Critical Approach to Science and Philosophy* (ed. M. Bunge), p. 266, London: Collier-Macmillan.
Eccles, J. C., 1964a, *The Physiology of Synapses*, Berlin: Springer.
Eccles, J. C., 1970, *Facing Reality*, Berlin: Springer.
Ekstein, R., 1966, in *Mind, Matter and Method* (ed. P. K. Feyerabend and G. Maxwell), p. 59, Minneapolis: Univ. of Minnesota Press.
Estes, W. K., Koch, S., McCorquodale, K., Meehl, P. E., Mueller, C. G., Schoenfeld, W. N., and Verplanck, W. S., 1954, *Modern Learning Theory*, New York: Appleton-Century-Crofts.
Ettlinger, G., and Wyke, M., 1961, *J. Neurol. Neurosurg. Psychiat.*, **24**, 254.
Eysenck, H. J., 1953, *Uses and Abuses of Psychology*, Harmondsworth: Penguin.
Eysenck, H. J. (ed.), 1960, *Handbook of Abnormal Psychology*, London: Pitman.
Eysenck, H. J., 1961, *Inquiry*, **5**, 1.

Farley, B. G., 1960, in *Self-Organizing Systems* (ed. M. C. Yovits and S. Cameron), p. 7, London: Pergamon.
Feigl, H., 1950, *Phil. Sci.*, **17**, 35.
Feigl, H., 1956, in *Minnesota Studies in the Philosophy of Science* (ed. H. Feigl and M. Scriven), vol. 1, p. 3, Minneapolis: Univ. of Minnesota Press.
Feigl, H., 1958, in *Minnesota Studies in the Philosophy of Science* (ed. H. Feigl, M. Scriven and G. Maxwell), vol. 2, p. 370, Minneapolis: Univ. of Minnesota Press.
Feigl, H., 1964, in *The Critical Approach to Science and Philosophy* (ed. M. Bunge), p. 45, London: Collier-Macmillan.
Fenichel, O., 1945, *The Psychoanalytic Theory of Neurosis*, New York: W. W. Norton & Co.
Fernandéz-Morán, H., 1971, *Industrial Research*, October, Chicago.
Fernandéz-Morán, H., 1972, *The University of Chicago Office of Public Information News Sheet*, 20 January.
Feyerabend, P. K., 1968, *Brit. J. Phil. Sci.*, **19**, 251.
Ficq, A., and Brachet, J., 1971, *Proc. Natl. Acad. Sci. U.S.*, **68**, 2774.
Fine, A. I., 1968, *Phil. Sci.*, **35**, 101.
Finkelstein, D., 1969, in *Boston Studies in the Philosophy of Science* (ed. R. S. Cohen and M. W. Wartofsky), vol. 5, p. 199, Dordrecht: D. Reidel Publ. Co.
Fodor, J. A., 1968, *Psychological Explanation*, New York: Random House.
Foster, L., 1969, *Brit. J. Phil. Sci.*, **20**, 259.
Fowler, R. H., 1936, *Statistical Mechanics* (2nd edition), Cambridge: Cambridge Univ. Press.
Fowler, R. H., and Guggenheim, E. A., 1939, *Statistical Thermodynamics*, Cambridge: Cambridge Univ. Press.
Frank, F. (ed.), 1973, *Water: A Comprehensive Treatise*, vols. 1–3, New York: Plenum Press.
Frank, H. S., 1970, *Science*, **169**, 635.
Frank, P., 1947, in *International Encyclopaedia of Unified Science*, **1**, no. 7, Chicago: Chicago Univ. Press.
Frank, P., 1957, *Philosophy of Science*, New York: Prentice-Hall.
Fraser Nicol, J., 1955, *J. Soc. Psych. Res.*, **38**, 71.
Freeman, G. L., 1948, *The Energetics of Human Behavior*, Ithaca: Cornell Univ. Press.
Fröhlich, H., 1969, *Brit. J. Phil. Sci.*, **20**, 167.

Gazzaniga, M. S., 1969, *The Bisected Brain*, New York: Appleton-Century-Crofts.

Gell-Mann, M., and Pais, A., 1955, *Phys. Rev.*, **97**, 1387.
Gellner, E., 1959, *Words and Things*, London: Gollancz.
Gendron, B., 1971, in *Boston Studies in the Philosophy of Science* (ed. C. Buck and R. S. Cohen), vol. 8, p. 483, Dordrecht: D. Reidel Publ. Co.
George, F. H., 1956, *Philosophy*, **31**, 244.
George, F. H., 1958, *Philosophy*, **33**, 57.
George, F. H., 1962, *Philosophy*, **37**, 62.
Geschwind, N., 1969, in *Boston Studies in the Philosophy of Science* (ed. R. S. Cohen and M. W. Wartofsky), vol. 4, p. 98, Dordrecht: D. Reidel Publ. Co.
Gibson, J. J., 1950, *The Perception of the Visual World*, Boston: Houghton Mifflin.
Glauber, R. J., 1970, in *Quantum Optics* (ed. S. M. Kay and A. Maitland), p. 53, New York: Academic Press.
Gluskov, V. M., 1961, *Usp. Mat. Nauk.*, **16**, 5.
Gödel, K., 1930a, *Monatschefte Math. Phys.*, **37**, 349.
Gödel, K., 1930b, *Monatshefte Math. Phys.*, **38**, 173.
Goldstein, H., 1950, *Classical Mechanics*, Reading, Mass.: Addison-Wesley.
Goldstein, K., 1940, *Human Nature in the Light of Psychology* (The William James Lectures), Cambridge, Mass: Harvard Univ. Press.
Golightly, C. L., 1952, *Phil. Sci.*, **19**, 225.
Good, I. J., 1968, *Brit. J. Phil. Sci.*, **19**, 357.
Goodman, N., 1965, *Fact, Fiction and Forecast* (2nd edition), Indianapolis: Bobbs-Merrill.
Grant, C. K., 1955, *Proc. 2nd Internat. Congress of the Internat. Union for the Philosophy of Science*, Neuchatel: Edition du Griffon.
Griffith, J. S., 1971, *Mathematical Neurobiology*, New York: Academic Press.

Haggard, H. W., 1934, *The Doctor in History*, New Haven: Yale Univ. Press.
Haken, H., 1970, in *Quantum Optics* (ed. S. M. Kay and A. Maitland), p. 201, New York: Academic Press.
Hamlyn, D. W., 1953, *Philosophy*, **28**, 132.
Hanson, W. H., 1971, *Brit. J. Phil. Sci.*, **22**, 9.
Harnwell, G. P., and Livingood, J. J., 1933, *Experimental Atomic Physics*, New York: McGraw-Hill.
Harré, R., 1970, *The Principles of Scientific Thinking*, London: Macmillan.
Harrison, W. A., 1970, *Solid State Theory*, New York: McGraw-Hill.
Hartmann, M., 1933, *Die Methodologischen Grundlagen der Biologie*, Leipzig: Meiner.
Hawkins, J. K., 1961, *Proc. Inst. Radio Engrs*, **49**, 31.
Hebb, D. O., 1954, in *Brain Mechanisms and Consciousness* (ed. J. F. Delafresnaye), p. 402, Oxford: Blackwell.
Hebb, D. O., 1959, *Brain*, **82**, 260.
Hebb, D. O., and Bindra, D., 1952, *Amer. J. Psychol.*, **7**, 569.
Hécaen, H., and Angelergues, R., 1963, *La Cécité psychique*, Paris: Masson.
Heisenberg, W., 1930, *The Physical Principles of the Quantum Theory*, Chicago: Chicago Univ. Press.
Heitler, W., 1945, *Elementary Wave Mechanics*, Oxford: Clarendon Press.
Heitler, W., 1954, *The Quantum Theory of Radiation* (3rd edition), Oxford: Clarendon Press.
Hempel, C. G., 1958, in *Minnesota Studies in the Philosophy of Science* (ed. H. Feigl, M. Scriven and G. Maxwell), vol. 2, p. 37, Minneapolis: Univ. of Minnesota Press.
Hempel, C. G., 1966, in *Mind and Cosmos* (ed. R. G. Colodny), Pittsburgh: Univ. of Pittsburgh Press.
Hempel, C. G., 1966a, *Philosophy of Natural Science*, Englewood Cliffs, N.J.: Prentice-Hall.
Hempel, C. G., 1967, *Aspects of Scientific Explanation*, New York: Appleton-Century-Croft.
Hempel, C. G., and Oppenheim, P., 1948, *Phil. Sci.*, **15**, 135.
Hesse, M. B., 1953, *Brit. J. Phil. Sci.*, **4**, 198.
Hesse, M. B., 1964, in *Quanta and Reality* (ed. D. Bohm), p. 49, New York: World Publ. Co.
Hesse, M. B., 1966, *Models and Analogies in Science*, Notre Dame, Indiana: Univ. of Notre Dame Press.
Hesse, M. B., 1968, *Brit. J. Phil. Sci.*, **19**, 176.

Hesse, M. B., 1969, *Brit. J. Phil. Sci.*, **20,** 13.
Hilgard, E. R., and Marquis, D. G., 1961, *Conditioning and Learning* (revised by G. A. Kimble), London: Methuen.
Hinde, R. E., 1966, *Animal Behavior*, New York: McGraw-Hill.
Hirst, R. J., 1959, *The Problems of Perception*, London: Allen & Unwin.
Hochberg, J. E., 1964, *Perception*, Englewood Cliffs, N.J.: Prentice-Hall.
Hochberg, J. E., 1971, in *Woodworth and Schlosberg's Experimental Psychology* (ed. J. W. Kling and L. A. Riggs), pp. 395, 475, London: Methuen.
Hodgkin, A. L., 1964, *The Conduction of Nerve Impulses*, Springfield, Illinois: Thomas.
Hogben, L., 1957, *Statistical Theory*, London: Allen & Unwin.
Hook, S. (ed.), 1959, *Psychoanalysis, Scientific Method, and Philosophy*, New York: New York Univ. Press.
Hull, C. L., 1943, *The Principles of Behavior*, New York: Appleton-Century-Croft.
Hull, C. L., 1950, *Psychol. Rev.*, **57,** 173.
Hull, C. L., 1951, *Essentials of Behavior*, New Haven: Yale Univ. Press.
Hull, C. L., 1952, *A Behavior System*, New Haven: Yale Univ. Press.
Hull, C. L., Hovland, C. I., Ross, R. T., Hall, M., Perkins, D. T., and Fitch, F. B., 1940, *Mathematico-deductive Theory of Rote Learning: a study in scientific methodology*, New Haven: Yale Univ. Press.
Huxley, J. S., 1942, *Evolution: the Modern Synthesis*, London: Allen & Unwin.

Jackson, J. D., 1958, *The Physics of Elementary Particles*, Princeton, N.J.: Princeton Univ. Press.
Jackson, L. F., 1958, *Brit. J. Psychol. Statist. Sect.*, **11,** 76.
Jacob, F., and Monod, J., 1963, in *Biological Organization at the Cellular and Supercellular Level* (ed. R. J. C. Harris), part 1, New York: Academic Press.
Jaeger, J. C., 1949, *An Introduction to the Laplace Transformation*, London: Methuen.
Jánossy, L., 1965, *Theory and Practice of the Evaluation of Measurements*, Oxford: Clarendon Press.
Jeffrey, R. C., 1971, in *Boston Studies in the Philosophy of Science* (ed. C. Buck and R. S. Cohen), vol. 8, p. 40, Dordrecht: D. Reidel Publ. Co.
Jeffreys, H., 1957, *Scientific Inference* (2nd edition), Cambridge: Cambridge Univ. Press.
Jeffreys, H., 1961, *Cartesian Tensors*, Cambridge: Cambridge Univ. Press.
Joos, G., 1947, *Theoretical Physics*, London: Blackie.

Kapp, R. O., 1951, *Mind, Life and Body*, London: Constable.
Kapp, R. O., 1955, *Facts and Faith: the Dual Nature of Reality* (Riddell Memorial Lectures, University of Durham), London: Oxford Univ. Press.
Kato, T., 1966, *Perturbation Theory for Linear Operators*, Berlin: Springer.
Kemble, E. C., 1937, *The Fundamental Principles of Quantum Mechanics*, New York: McGraw-Hill.
Kemeny, J. G., 1953, *Phil. Rev.*, **62,** 93.
Kemeny, J. G., 1953a, *Phil. Rev.*, **62,** 391.
Kleene, S. C., 1952, *Introduction to Metamathematics*, Amsterdam: North-Holland Publ. Co.
Kleinschmidt, A. K., Lang, D., Jacherts, D., and Zahn, R. R., 1962, *Biochim. biophys. acta*, **61,** 857.
Kline, P., 1971, *Fact and Fantasy in Freudian Theory*, London: Methuen.
Kling, J. W., and Riggs, L. A. (ed.), 1971, *Woodworth and Schlosberg's Experimental Psychology*, London: Methuen.
Koch, S., 1954, in *Modern Learning Theory* (ed. W. K. Estes *et al.*), p. 1, New York: Appleton-Century-Crofts.
Koestler, A., 1972, *The Roots of Coincidence*, London: Hutchinson.
Köhler, W., 1938, *The Place of Value in a World of Facts*, New York: Liveright Publ. Corp.
Körner, S., 1955, *Kant*, Harmondsworth: Penguin.
Körner, S., 1966, *Experience and Theory*, London: Routledge & Kegan Paul.

Kraft, V., 1966, in *Mind, Matter and Method* (ed. P. K. Feyerabend and G. Maxwell), p. 306, Minneapolis: Univ. of Minnesota Press.

Kramers, H. A., 1938, *Die Grundlagen der Quantentheorie*, Leipzig: Akademische Verlagsgesellschaft (also translated into English as *Quantum Mechanics*, Amsterdam: North-Holland Publ. Co.).

Lakatos, I. (ed.), 1968, *The Problems of Inductive Logic*, Amsterdam: North-Holland Publ. Co.

Lange, J., 1936, in *Handbuch der Neurologie* (ed. O. Bumke and O. Foerster), vol. 6, p. 807, Berlin: Springer.

Lashley, K. S., 1950, *Symp. Soc. Exp. Biol.*, **4,** 454.

Lashley, K. S., 1951, in *Cerebral Mechanisms in Behavior* (ed. L. A. Jeffreys), p. 230, New York: Wiley.

Leder, P., and Nirenberg, M. W., 1964a, *Proc. Natl. Acad. Sci. U.S.*, **52,** 1521.

Leder, P., and Nirenberg, M. W., 1964b, *Proc. Natl. Acad. Sci. U.S.*, **52,** 420.

Le Pichon, X., Francheteau, J. and Bonnin, J., 1973, *Plate Tectonics*, Amsterdam: Elsevier.

Lewis, D., 1970, *Phil. Sci.*, **37,** 100.

Lewontin, R. C., 1972, *Nature*, **236,** 181.

Lissauer, H., 1889, *Arch. Psychiat. Nervenkr.*, **21,** 222.

Litwin, S. D., Lin, P. K., Hatteroth, T. H., and Cleve, H., 1973, *Nature New Biology*, **246,** 179.

Lomnitz, C., 1973, *Global Tectonics and Earthquake Risk*, Amsterdam: Elsevier.

Lorenz, K. Z., 1950, *Symp. Soc. Exp. Biol.*, **4,** 221.

Lorenz, K. Z., 1966, *Evolution and Modification of Behaviour*, London: Methuen.

Lucas, J. R., 1961, *Philosophy*, **36,** 112.

Lucas, J. R., 1968a, *Monist*, **52,** 145.

Lucas, J. R., 1968b, *Brit. J. Phil. Sci.*, **19,** 155.

Lüders, G., 1961, in *Lectures on Field Theory and the Many-Body Problem* (ed. E. R. Caianiello), p. 1, New York: Academic Press.

MacKay, D. M., 1952, *Aristot. Soc. Suppl.*, **26,** 61.

McCorquodale, K., and Meehl, P. E., 1948, *Psychol. Rev.*, **55,** 95.

McDougall, W., 1934, *Modern Materialism and Emergent Evolution* (2nd edition), London: Methuen.

McWeeny, R., and Sutcliffe, B. T., 1969, *Methods of Molecular Quantum Mechanics*, London: Academic Press.

Macrae, D., and Trolle, E., 1956, *Brain*, **79,** 94.

Madden, E. H., 1952, *Phil. Sci.*, **19,** 228.

Margenau, H., and Cohen, L., 1967, in *Studies in the Foundations, Methodology and Philosophy of Science* (ed. M. Bunge), vol. 2, p. 71, Berlin: Springer.

Marx, M. H., 1952, *Psychological Theory*, New York: Macmillan.

Masterman, M., 1957, in *British Philosophy in the Mid-Century* (ed. C. A. Mace), London: Allen & Unwin.

Masters, M., and Broda, P., 1971, *Nature New Biology*, **232,** 137.

Mayr, E., 1961, *Science*, **134,** 1501.

Mays, W., 1952, *Philosophy*, **27,** 148.

Medawar, P. B., 1961, *Mind*, **70,** 99.

Mercier, A., 1970, in *Physics, Logic and History* (ed. W. Yourgrau and A. D. Breck), p. 39, New York: Plenum Press.

Meyer, H., 1951, *Phil. Sci.*, **18,** 111.

Midgley, G. C., 1955, *Aristot. Soc. Suppl.*, **29,** 185.

Mill, J. S., 1865, *Examination of Sir William Hamilton's Philosophy*, London: Longman.

Miller, G. A., Galanter, E., and Pribram, K. H., 1960, *Plans and the Structure of Behavior*, New York: Henry Holt.

Miller, O. L., 1973, *Scientific American*, **228,** no. 3, 34.

Minsky, M. L., 1959, *Mechanisation of Thought Processes*, National Physical Laboratory Symposium no. 10, London: Her Majesty's Stationery Office.

Minsky, M. L., 1961, *Proc. Inst. Radio Engrs*, **49,** 8.

Minsky, M. L., 1967, *Computation: Finite and Infinite Machines*, Englewood Cliffs, N.J.: Prentice-Hall.
Moore, G. E., 1922, *Philosophical Studies*, London: Routledge & Kegan Paul.
Moore, G. E., 1953, *Some Main Problems of Philosophy*, London: Allen & Unwin.
Moore, G. E., 1959, *Philosophical Papers*, London: Allen & Unwin.
Morse, P. M., 1968, *Theoretical Acoustics*, New York: McGraw-Hill.
Moyal, J. E., 1949a, *J. Roy. Statist. Soc. Series B*, **11,** 150.
Moyal, J. E., 1949b, *Proc. Camb. Phil. Soc.*, **45,** 99.
Moyal, J. E., 1949c, *Philosophy*, **24,** 310.
Mundle, C. W. K., 1970, *A Critique of Linguistic Philosophy*, Oxford: Clarendon Press.
Murphy, G., 1953, *Proc. Soc. Psych. Res.*, **50,** 26.

Needham, J., 1927, *Man a Machine*, London: Kegan Paul, Trench, Trubner & Co.
Needham, J., 1942, *Biochemistry and Morphogenesis*, Cambridge: Cambridge Univ. Press.
Neisser, U., 1967, *Cognitive Psychology*, New York: Appleton-Century-Crofts (Meredith Publ. Co.).
Neurath, O., 1931, *Monist*, **41,** 618.
Nilsson, N. J., 1965, *Learning Machines*, New York: McGraw-Hill.

Orgel, L. E., 1959, *Rev. Mod. Phys.*, **31,** 100.

Pauli, W., 1933, in *Handbuch der Physik* (ed. H. Geiger and K. Scheel) vol. 24, part 1, p. 83, Berlin: Springer.
Pavlidis, T., 1971, *J. Theoret. Biol.*, **33,** 319.
Pecker, J. C., Roberts, A. P. and Vigier, J. P., 1972, *Nature*, **237,** 227.
Peierls, R. E., 1955, *Quantum Theory of Solids*, Oxford: Clarendon Press.
Penfield, W., and Rasmussen, T., 1950, *The Cerebral Cortex of Man*, New York: Macmillan.
Penrose, L. S., 1957, *The Listener*, **57,** 748.
Perutz, F. M., 1969, *Proc. Roy. Soc. B*, **173,** 113.
Perutz, F. M., 1970, *Nature*, **228,** 726.
Peters, R. S., 1958, *The Concept of Motivation*, London: Routledge & Kegan Paul.
Peters, R. S. (ed.), 1962, *Brett's History of Psychology*, London: Allen & Unwin.
Pfenninger, K., Sandri, C., Akert, K., and Eugster, C. H., 1969, *Brain Res.*, **12,** 10.
Philbrick, F. A., and Holmyard, E. J., 1945, *A Textbook of Theoretical and Inorganic Chemistry*, London: J. M. Dent & Sons.
Pittendrigh, C. S., 1961, *Cold Spring Harb. Symp. Quant. Biol.*, **25,** 159.
Poincaré, H., 1898, *Monist*, **9,** 42.
Poincaré, H., 1905, *Science and Hypothesis*, London: Walter Scott.
Polanyi, M., 1961, *The Observer*, London, 21 May, p. 18.
Polanyi, M., 1968, *Science*, **160,** 1308.
Pole, D., 1958, *The Later Philosophy of Wittgenstein*, London: Univ. of London, The Athlone Press.
Popper, K. R., 1960, *Proc. Brit. Acad.*, **46,** 39.
Popper, K. R., 1960a, *The Logic of Scientific Discovery*, London: Hutchinson.
Popper, K. R., 1963, *Synthese*, **15,** 167.
Popper, K. R., 1963a, *Conjectures and Refutations: The Growth of Scientific Knowledge*, London: Routledge & Kegan Paul.
Popper, K. R., 1967, in *Studies in the Foundations, Methodology and Philosophy of Science* (ed. M. Bunge), vol. 2, p. 7, Berlin: Springer.
Popper, K. R., 1970, in *Physics, Logic, and History* (ed. W. Yourgrau and A. D. Breck), p. 1, New York: Plenum Press.
Pribram, K. H., 1971, *Languages of the Brain*, Englewood Cliffs, N.J.: Prentice-Hall.
Price, H. H., 1939, *Proc. Soc. Psych. Res.*, **45,** 307.
Pullman, B., and Pullman, A., 1963, *Quantum Biochemistry*, New York: Interscience.
Putnam, H., 1969, in *Boston Studies in the Philosophy of Science* (ed. R. S. Cohen and M. W. Wartofsky), vol. 5, p. 216, Dordrecht: D. Reidel Publ. Co.

Quine, W. V. O., 1953, *From a Logical Point of View*, Cambridge, Mass.: Harvard Univ. Press.
Quine, W. V. O., 1961, *Word and Object*, New York: Wiley.

Rashevsky, N., 1948, *Mathematical Biophysics*, 2 vols., New York: Dover.
Ravetz, J. R., 1971, *Scientific Knowledge and its Social Problems*, Oxford: Clarendon Press.
Rawcliffe, D. H., 1952, *The Psychology of the Occult*, London: Derrick Ridgeway.
Reichenbach, H., 1949, *The Theory of Probability*, Berkeley and Los Angeles: Univ. of California Press.
Reichenbach, H., 1961, *Experience and Prediction* (1938; reissued 1961), Chicago: Chicago Univ. Press.
Renner, B., 1968, *Current Algebras and their Applications*, Oxford: Pergamon.
Rock, I., 1966, *The Nature of Perceptual Adaptation*, New York: Basic Books Inc.
Rozeboom, W., 1956, *Psychol. Rev.*, **63,** 249.
Rozeboom, W., 1968, *Phil. Sci.*, **35,** 134.
Rozeboom, W., 1971, in *Boston Studies in the Philosophy of Science* (ed. R. C. Buck and R. S. Cohen), vol. 8, p. 342, Dordrecht: D. Reidel Publ. Co.
Russell, B., 1953, *Mysticism and Logic*, Harmondsworth: Penguin.
Ryle, G., 1949, *The Concept of Mind*, London: Hutchinson Univ. Library.
Ryle, G., 1951, in *The Physical Basis of Mind* (ed. P. Laslett), p. 75, Oxford: Blackwell.

Salmon, W. C., 1966, *The Foundations of Scientific Inference*, Pittsburgh: Univ. of Pittsburgh Press.
Sanger, F., and Tuppy, H., 1951a, *Biochem. J.*, **49,** 463.
Sanger, F., and Tuppy, H., 1951b, *Biochem. J.*, **49,** 481.
Scheffler, I., 1963, *The Anatomy of Inquiry*, New York: Knopf.
Scheffler, I., 1967, *Science and Subjectivity*, Indianapolis: Bobbs-Merrill.
Schlick, M., 1925, *Allgemeine Erkenntnislehre*, Berlin: Springer.
Schlick, M., 1935, *Synthese*, **10,** 5.
Schlick, M., 1948, *Philosophy of Nature*, New York: Philosophical Library.
Schweber, S. S., 1964, *An Introduction to Relativistic Quantum Field Theory*, New York: Harper & Row.
Scriven, M., 1959, *Science*, **130,** 477.
Scriven, M., 1966, in *Mind, Matter and Method* (ed. P. K. Feyerabend and G. Maxwell), p. 191, Minneapolis: Univ. of Minnesota Press.
Sellars, W., 1948, *Philosophy and Phenomenological Research*, **8,** 601 (reprinted in *Readings in Philosophical Analysis* (ed. H. Feigl and W. Sellars), p. 424, New York: Appleton-Century-Crofts, 1949).
Sellars, W., 1954, *Phil. Sci.*, **21,** 204.
Seward, J. P., 1954, *Psychol. Rev.*, **61,** 145.
Sinnot, E. W., Dunn, L. C., and Dobzhansky, Th. 1958, *Principles of Genetics* (5th edition), New York: McGraw-Hill.
Skinner, B. F., 1947, in *Current Trends in Psychology* (ed. W. Dennis), Pittsburgh: Pittsburgh Univ. Press.
Skinner, B. F., 1953, *Science and Human Behavior*, New York: Macmillan.
Smuts, J. C., 1926, *Holism and Evolution*, London: Macmillan.
Sommerfeld, A., 1939, *Atombau und Spektrallinien: Wellenmechanischer Ergänzungsband* (2nd edition), Braunschweig: Vieweg und Sohn.
Sommerhof, G., 1950, *Analytical Biology*, London: Oxford Univ. Press.
Spence, K. W., 1944, *Psychol. Rev.*, **51,** 47.
Spence, K. W., 1956, *Behavior Theory and Conditioning* (Silliman Memorial Lectures), New Haven: Yale Univ. Press.
Sperry, R. W., 1963, *Proc. Natl. Acad. Sci. U.S.*, **50,** 703.
Sperry, R. W., 1969, *Psychol. Rev.*, **76,** 532.
Sperry, R. W., 1970, *Psychol. Rev.*, **77,** 585.
Spitz, H. H., 1958, *Psychol. Bull.*, **55,** 1.
Stratton, J. A., 1941, *Electromagnetic Theory*, New York: McGraw-Hill.

Suppes, P., 1966, in *Aspects of Inductive Logic* (ed. J. Hintikka and P. Suppes), Amsterdam: North-Holland Publ. Co.
Synge, J. L., 1960, in *Handbuch der Physik* (2nd edition) (ed. S. Flügge), vol. 3, part 1, Berlin: Springer.
Synge, J. L., and Griffith, B. A., 1942, *Principles of Mechanics*, New York: McGraw-Hill.

Teilhard de Chardin, P., 1960, *The Phenomenon of Man*, London: Collins.
Temin, H. M., and Mizutani, S., 1970, *Nature*, **226,** 1211.
Teuber, H. L., 1960, in *Handbook of Physiology and Neurophysiology* (ed. J. Field and H. W. Magoun), § 1, p. 3, Washington D.C.: American Physiological Society.
Theimer, O., Wassermann, G. D., and Wolf, E., 1952, *Proc. Roy. Soc. A*, **212,** 426.
Thompson, D'Arcy, W., 1942, *On Growth and Form*, Cambridge: Cambridge Univ. Press.
Tinbergen, N., 1951, *The Study of Instinct*, London: Oxford Univ. Press.
Tolman, E. C., 1932, *Purposive Behavior in Animals and Men*, New York: Appleton-Century-Crofts.
Tolman, E. C., 1938, *Psychol. Rev.*, **45,** 1.
Tolman, E. C., 1945, *Science*, **101,** 160.
Tolman, E. C., 1949, *Psychol. Rev.*, **56,** 357.
Tolman, E. C., 1955, *Psychol. Rev.*, **62,** 315.
Tolman, R. C., 1938, *The Principles of Statistical Mechanics*, Oxford: Clarendon Press.
Toulmin, S., 1953, *The Philosophy of Science*, London: Hutchinson.
Toulmin, S., 1971, in *Boston Studies in the Philosophy of Science* (ed. R. C. Buck and R. S. Cohen), vol. 8, p. 552, Dordrecht: D. Reidel Publ. Co.
Tryon, E. P., 1973, *Nature*, **246,** 396.
Turing, A. M., 1937a, *Proc. Lond. Math. Soc. Ser. 2*, **42,** 230.
Turing, A. M., 1937b, *Proc. Lond. Math. Soc. Ser. 2*, **43,** 544.
Turing, A. M., 1950, in *Minds and Machines* (ed. A. R. Anderson), p. 4, Englewood Cliffs, N.J.: Prentice-Hall.
Turing, A. M., 1952, *Phil. Trans. Roy. Soc. B*, **237,** 37.
Tyrrell, G. N., 1947, *The Personality of Man*, Harmondsworth: Penguin.

van der Waerden, B. L., 1932, *Die Gruppentheoretische Methode in der Quantenmechanik*, Berlin: Springer.
Vernon, M. D., 1952, *A Further Study of Visual Perception*, Cambridge: Cambridge Univ. Press.
Vernon, M. D., 1962, *The Psychology of Perception*, Harmondsworth: Penguin.
Vernon, M. D., 1970, *Perception through Experience*, London: Methuen.
Verplanck, W. S., 1954, in *Modern Learning Theory* (ed. W. K. Estes *et al.*), p. 267, New York: Appleton-Century-Crofts.
Visconti, A., 1969, *Quantum Field Theory*, 2 vols., London: Pergamon.
von Bertalanffy, L., 1950a, *Science*, **111,** 23.
von Bertalanffy, L., 1950b, *Brit. J. Phil. Sci.*, **1,** 134.
von Mises, R., 1919 *Mathematische Zeitschr*, **5,** 53.
von Mises, R., 1931, *Wahrscheinlichkeitsrechnung*, Leipzig: Deuticke.
von Wright, G. H., 1957, *The Logical Problems of Induction* (2nd edition), Oxford: Blackwell.

Waddington, C. H., 1956, *Principles of Embryology*, London: Allen & Unwin.
Waddington, C. H., 1957, *The Strategy of the Genes: a Discussion of some Aspects of Theoretical Biology* (with an appendix by H. Kacser), London: Allen & Unwin.
Waddington, C. H., 1966, *Principles of Development and Differentiation*, London: Collier-Macmillan.
Wald, A., 1952, *On the Principles of Statistical Inference*, Notre Dame, Indiana: Notre Dame Univ. Press.
Walshe, F. M. R., 1953, *Brain*, **76,** 1.
Walter, W. G., 1953, *The Living Brain*, London: Gerald Duckworth.
Wang, H., 1963, *A Survey of Mathematical Logic*, Amsterdam: North-Holland Publ. Co.

Wassermann, G. D., 1946, *Phil. Mag. Ser. 7*, **37**, 563.
Wassermann, G. D., 1947, *Proc. Camb. Phil. Soc.*, **44**, 251.
Wassermann, G. D., 1949, *Proc. Camb. Phil. Soc.*, **46**, 206.
Wassermann, G. D., 1955, *Brit. J. Phil. Sci.*, **6**, 122.
Wassermann, G. D., 1952, *Quart. J. Mech. Appl. Math.*, **5**, 466.
Wassermann, G. D., 1972, *Molecular Control of Cell Differentiation and Morphogenesis*, New York: Marcel Dekker Inc.
Wassermann, G. D., 1973, *Nature New Biology*, **245**, 163.
Wassermann, G. D., 1974, *Psychobiological and Neuroscientific Theories*, New York: Marcel Dekker Inc. (in the press).
Wathen-Dunn, W. (ed.), 1967, *Models for the Perception of Speech and Visual Form*, Cambridge, Mass.: MIT Press.
Watson, J. B., 1919, *Psychology from the Standpoint of a Behaviorist*, Philadelphia: Lippincott.
Weatherburn, C. E., 1952, *Mathematical Statistics*, Cambridge: Cambridge Univ. Press.
Weinberg, J. R., 1936, *An Examination of Logical Positivism*, London: Kegan Paul, Trench, Trubner & Co.
Wentzel, G., 1949, *Quantum Theory of Fields*, New York: Interscience Publishers.
Weiss, P. A., 1971, in *The Neurosciences: 2nd Study Program* (ed. F. O. Schmitt), p. 53, New York: Rockefeller Univ. Press.
Wertheimer, M., 1959, *Productive Thinking*, London: Tavistock.
Whitehead, S., 1951, *Dielectric Breakdown of Solids*, Oxford: Clarendon Press.
Whitley, C. H., 1962, *Philosophy*, **37**, 61.
Whittaker, E., 1951, *A History of the Theories of Aether and Electricity* (*The Classical Theories*), Edinburgh: Thomas Nelson.
Wiener, N., 1948, *Cybernetics, or Control and Communication in the Animal and the Machine* (1st edition), New York: Wiley.
Wiener, N., 1961, *Cybernetics, or Control and Communication in the Animal and the Machine* (2nd edition), New York: Wiley.
Wilson, A. H., 1958, *The Theory of Metals* (2nd edition), Cambridge: Cambridge Univ. Press.
Wilson, W., 1940, *Theoretical Physics*, vol. 3, London: Methuen.
Winterstein, H., 1928, *Kausalität und Vitalismus vom Standpunkt der Denkökonomie*, Berlin: Springer.
Wisdom, J. O., 1964, in *The Critical Approach to Science and Philosophy* (ed. M. Bunge), p. 116, London: Collier-Macmillan.
Wittgenstein, L., 1953, *Philosophical Investigations*, Oxford: Blackwell.
Woodger, J. H., 1947, in *International Encyclopaedia of Unified Science*, **2**, no. 5, Chicago: Chicago Univ. Press.
Woodger, J. H., 1952, *Biology and Language*, Cambridge: Cambridge Univ. Press.
Woodger, J. H., 1956, *Physics, Psychology and Medicine*, Cambridge: Cambridge Univ. Press.
Woodworth, R. S., and Schlosberg, H., 1955, *Experimental Psychology* (2nd edition), London: Methuen.

Yourgrau, W., 1964, in *The Critical Approach to Science and Philosophy* (ed. M. Bunge), p. 360, London: Collier-Macmillan.
Yourgrau, W., 1970, in *Physics, Logic and History* (ed. W. Yourgrau and A. D. Breck), p. 125, New York: Plenum Press.

Zusne, L., 1970, *Visual Perception of Form*, New York: Academic Press.

AUTHOR INDEX

SUBJECT INDEX